E-MRS Spring Meeting 2014 Symposium D: Phonons and Fluctuations in Low Dimensional Structures

IOP Conference Series: Materials Science and Engineering Volume 68

Lille, France
26 – 30 May 2014

ISBN: 978-1-63439-879-4
ISSN: 1757-8981

Printed from e-media with permission by:

Curran Associates, Inc.
57 Morehouse Lane
Red Hook, NY 12571

Some format issues inherent in the e-media version may also appear in this print version.

Printed by Curran Associates, Inc. (2015)

Institute of Physics
Dirac House, Temple Back
Bristol BS1 6BE UK

Phone: 44 1 17 929 7481
Fax: 44 1 17 920 0979

techtracking@iop.org

Additional copies of this publication are available from:

Curran Associates, Inc.
57 Morehouse Lane
Red Hook, NY 12571 USA
Phone: 845-758-0400
Fax: 845-758-2634
Email: curran@proceedings.com
Web: www.proceedings.com

TABLE OF CONTENTS

IOP Conf. Series: Materials Science and Engineering **68** (2014) 011001 doi:10.1088/1757-899X/68/1/011001

The E-MRS 2014 Spring meeting, held from 26-30th May 2014 in Lille included the Symposium D entitled "Phonons and Fluctuations in Low Dimensional Structures", the first edition of its kind.
The symposium was organised in response to the increasing interest in the study of phonons in the context of advances in condensed matter physics, electronics, experimental methods and theory and, in particular, the transfer of energy across atomic interfaces and the propagation of energy in the nm-scale.

Steering heat by light or vice versa and examining nano-scale energy conversion (as in thermoelectricity and harvesting e.g. in biological systems) are two aspects that share the underlying science of energy processes across atomic interfaces and energy propagation in the nanoscale and or in confined systems. The nanometer scale defies several of the bulk relationships as confinement of electrons and phonons, locality and non-equilibrium become increasingly important. The propagation of phonons as energy carriers impacts not only heat transfer, but also the very concept and handling of temperature in non-equilibrium and highly localised conditions. Much of the needed progress depends on the materials studied and this symposium targeted the interface material aspects as well as the emerging concepts to advance in this field.
The symposium had its origins in a series of meetings and seminars including: (1) the first Phonon Engineering Workshop, funded by Catalan Institute for Research and Advanced Studies (ICREA), the then MICINN, the CNRS, VTT, and several EU projects, held in Saint Feliu de Guixols (Girona, Spain) from 24th to 27th of May 2010 with 65 participants from Europe, the USA and Japan; (2) the first Phonons and Fluctuations workshop, held in Paris on 8th and 9th November 2010, supported by French, Spanish and Finnish national projects and EU projects, attended by about 50 researchers; (3) the second Phonon and Fluctuations workshop, held in Paris on 8th and 9th September 2011, attended by 55 researchers and (4) the 3rd Workshop on Phonons & Fluctuations, held in Saint Feliux de Guixols (Girona, Spain) during 21 to 24th May 2012 attended by 65 participants from Europe and the USA.

These papers in this proceeding are examples of the work presented at the symposium. They represent the tip of the iceberg, as the symposium attracted over 100 abstracts. The meeting room was usually full with an audience varying between 40 and 100 participants. The plenary presentation was given by Prof. Gang Chen (MIT) on "Ballistic and Coherent Phonon Heat Conduction in Bulk Materials and Nanostructures", which was warmly welcome by an eager and highly motivated audience. The invited speakers were: Prof. Thomas Dehoux (U. Bordeaux), Dr S. Chung (U. New South Wales, Australia), Prof. A. Goni (CSIC-ICMAB), Prof. Giuliano Benenti (U. Insubria), Dr. Davide Donadio (Max Planck Institute for Polymer Research, Mainz), Prof. George Fytas (University of Crete), Prof. Dr. Tobias Kippenberg (EPFL, Switzerland), Prof. Bernard Perrin (INSP, Paris), Prof. Gyaneshwar P. Srivastava, U. Exeter) and Prof. Dr. Achim Kittel (U. Oldenburg). The organisers are very grateful to them for supporting the symposium and sharing their latest research results with the symposium participants.

The symposium organisers recognised the participation of students and awarded prizes to the two Best Student Presentations, which went to Valeria Lacatena (IEMN, Lille) with an invited presentation entitled "Efficient reduction of thermal conductivity in silicon using phononic-engineered membranes" and to Yan Qing Liu (Institute Neel, Grenoble) who presented the talk entitled "Sensitive 3-omega measurements of epitaxial thermoelectric thin films". The poster session had about fifty posters and the four best poster prizes went to: Konstanze Hahn et al. (U. Cagliari) poster title "Determination of Thermal conductivity in (nanostructured) SiGe materials", Florian Doehring et al. (U. Goettingen) poster title "Phonon blocking in Multilayers produced by Pulsed Laser Deposition", Jordi Gomis-Bresco et al. (ICN2) poster " A 1D PhoXonic Crystal", Barcelona and Benjamin J Robinson et al. (U. Lancaster) poster "Scanning Thermal microscopy studies of 2D materials".

The symposium organisers are grateful to the Scientific Committee members, Prof. Bahram Djafari-Rouhani (France), Prof. Dr. Thomas Dekorsy (Germany), Prof. Anthony Kent (UK), Prof. Fabio Marchesoni (Italy), Dr. Natalio Mingo (France), Prof. Pascal Ruello, (France) and Prof. Javier Viejo-Rodriguez (Spain), for their help with all aspects of evaluation of the scientific level of the presentations in the symposium.

The symposium was sponsored by the FP7 ICT FET Open Coordination Action EUPHONON (GA. 612086) and by the CNRS GDR "Thermal Nanosciences and NanoEngineering". The symposium organisers express their sincere thanks to the staff of the E-MRS for continuous support and timely advice in all organisational aspects. We are indebted to Dr. Erwan Guillotel (ICN2) for his assistance with the organisation of the symposium.

Jouni Ahopelto
VTT, Espoo, Finland

Clivia M. Sotomayor Torres
ICN2 & ICREA, Barcelona, Spain

Sebastian Volz
Ecole Centrale Paris and CNRS, Paris, France

Symposium Organisers.

October 2014.

Peer review statement

All papers published in this volume of *IOP Conference Series: Materials Science and Engineering* have been peer reviewed through processes administered by the proceedings Editors. Reviews were conducted by expert referees to the professional and scientific standards expected of a proceedings journal published by IOP Publishing.

EMRS Spring 2014 D IOP Publishing
IOP Conf. Series: Materials Science and Engineering **68** (2014) 012001 doi:10.1088/1757-899X/68/1/012001

Measurement of phonon damping by nanostructures

A. Al Mohtar[1,2], A. Bruyant[1], S. Kostcheev[1], J. Vaillant[1], A. Khoury[2], M. Kazan[3]

(1) UTT , ICD-LNIO and STMR- UMR6279 CNRS, 12 rue Marie Curie - CS 42060, 10004 Troyes - France, (2) LPA, Faculty of Sciences II, EDST, Lebanese University, P.O. Box 90239, Jdeidet 1202, Lebanon, (3) Department of Physics, American University of Beirut, P.O. Box 11-0236, Riad El-Solh, Beirut 1107-2020, Lebanon

E-mail: mk140@aub.edu.lb

Abstract. The understanding of phonon lifetime and scattering rates is attracting an increasing interest due to the major role of phonon in thermal and electrical conductivity which are key properties for technological applications. The infrared complex dielectric function of a crystal is determined by the harmonic characteristics of the phonon together with the intrinsic and extrinsic phonon scattering rates. In order to investigate the interplay between the phonon intrinsic scattering and the scattering of the phonon by a nanostructured surface, infrared reflectivity measurements from SiC nano-pyramids on SiC substrate have been analysed using a Kramers-Kronig conversion technique to deduce the infrared complex dielectric function. Then, the real and imaginary parts of the dielectric function were fitted simultaneously by using a theoretical model for the dielectric constant that considers frequency-dependent phonon damping at the center of the Brillouin zone. It has been found that surface nanostructuring strongly enhances the overall scattering rate of the phonon at the Brillouin zone center.

1. Introduction

Nanostructured SiC has been recently receiving a lot of attention from the materials and device community. Due to its wide potential applications ranging from microelectronics, optoelectronics, high temperature semiconductor nanocomposites, photocatalysts, electron field emitting devices to biomedical engineering and functional ceramics. All said applications rely on the unique attractive properties of SiC, including high temperature material stability, high thermal conductivity, low thermal expansion coefficient and thermal shock endurance [1–4]. As well as, the high mechanical strength of SiC enables its use in harsh environment. In order to better use and control the properties of the nanostructured SiC, a quantitative understanding of phonon lifetime and scattering rates is mandatory [5] since the phonon response controls the thermal and electrical conductivity in the infrared spectral range. For this purpose SiC nanotructures of diameter ranging between 70 nm and 100 nm were fabricated on SiC. In this contribution the complex infrared dielectric function of nanostructured SiC was calculated and fitted to experimental data. The permittivity was calculated by a model that considers the material as coupled damped oscillators. Perturbation theories were applied to determine the anharmonic components of interatomic forces associated with interchanges of energies between phonon modes as function of frequency and temperature [6]. Kramers-Kronig conversion technique was employed to deduce the infrared complex dielectric function [7].

2. Experimental Details

We have purchased good-crystalline quality SiC wafers from Cree, Inc. The sample was cut into two pieces, where the nanostructures were fabricated on one of them. The pattern of the structures was drawn by electron-beam lithography, preceded by the deposition of positive tone PMMA resist, followed by lift-off procedure. An Al layer was then deposited as a mask for Reactive Ion Etching (RIE) of the patterned SiC substrate. RIE was operated with the use of inductively coupled plasma in a mixture of 60% SF6 and 40% O2 at 3 mTorr pressure. The residual mask was removed in aqueous solution of KOH.

The reflectivity from each sample is normalized to that of a gold coated mirror. The measurements were conducted at room temperature in the 400-1500 cm^{-1} spectral range. Near normal incidence angle was fixed with uncertainty of 15°. The FTIR setup is Nicolet 4700 spectrometer from the Thermo Electron Corporation with a DTGS detector and a KBr beam-splitter.

3. Results and Discussion

Figure (1) shows the fabricated nanostructured SiC with diameter ranging from 70 nm to 100 nm. We investigated the surface phonons in SiC and the effect of the structuring on their lifetime. We measured the intact surface SiC reflectivity and that of the surface with nanostructures. Figure (2) is the unpolarized infrared reflectivity measurements made on SiC unmodified wafer and on the fabricated nanostructured SiC. A clear difference is noticed between the two reflectance spectra. This difference can not be due to light diffusion because the size of the nanostructures is about two orders of magnitude smaller than the light wavelength.

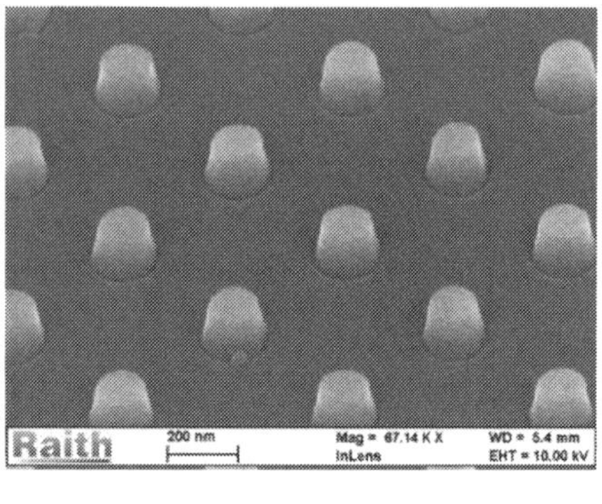

Figure 1: SEM image of SiC Nanostructures on SiC substrate.

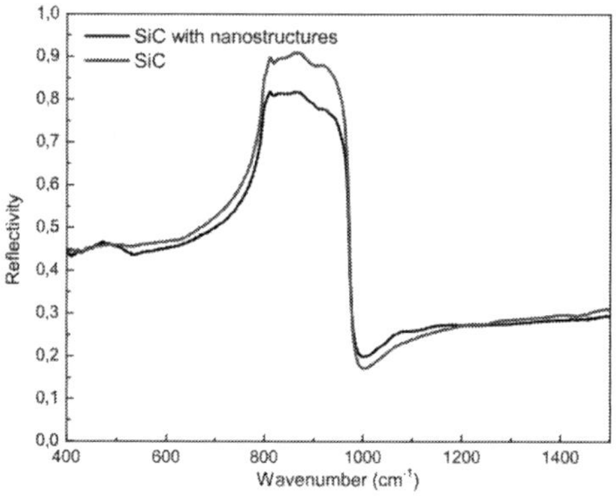

Figure 2: Reflectivity measurements on SiC wafer and nanostructured SiC.

The reflectivity measurements have been converted into real and imaginary parts of the dielectric functions. These functions are fitted simultaneously with a Lorentz model that considers the material as coupled oscillators with frequency-dependent damping constants [6]. Figure (3) and Figure (4) show the obtained results. The data from the nanostructured SiC is plotted along with that of the intact SiC. We see in symbols the experimental results and in solid lines the simulated ones. The best fit is mostly affected by considering the 3-phonon decays, which demonstrates that at room temperature the 3-phonon decay processes are the dominant mechanism.

A remarkable difference in the permittivity was observed due to the introduction of the nanostructures. From the fit we deduce that the presence of nanostructure has increased significantly the phonon damping. Figure (5) compares the damping constants of the two

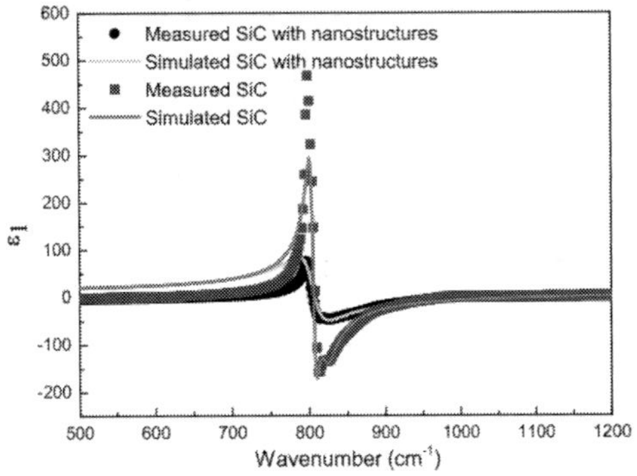

Figure 3: Real part of the dielectric function for SiC (blue), and nanostructured SiC (black). The symbols correspond to the measured data, the smooth line to the fitted one.

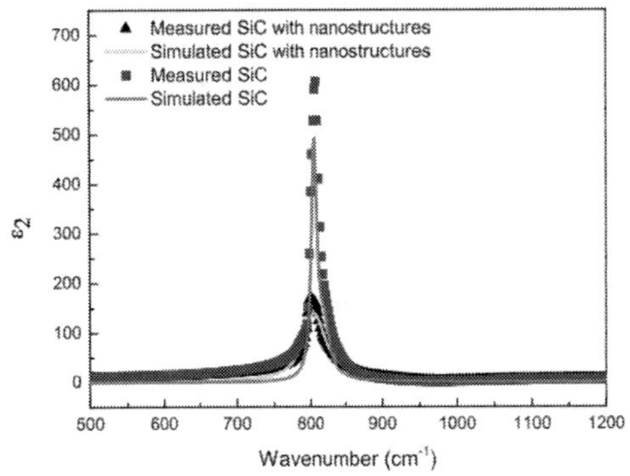

Figure 4: Imaginary part of the permittivity for SiC (blue), and nanostructured SiC (black). The symbols correspond to the measured data, the smooth line to the fitted one.

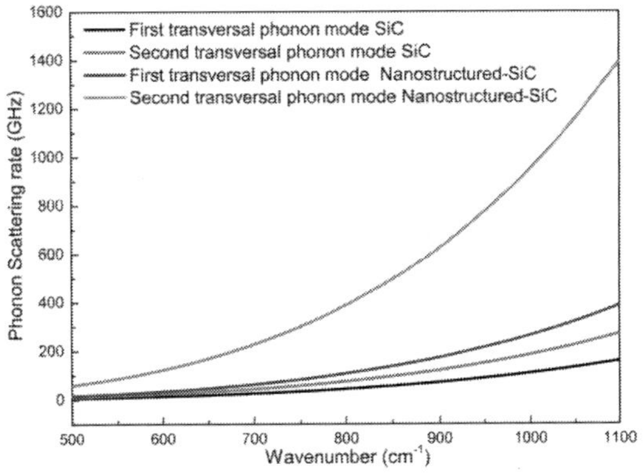

Figure 5: Calculated phonon scattering rate from the experimental fit.

investigated samples. We found that the damping constants related to the intact sample is ranging from 6 times to an order of magnitude smaller than that of the surface with nanostructures depending on the frequency range. Since the two measured samples come from the same epitaxial layer, we attribute the enhancement of the scattering rate to collisions between the surface phonons and the structures. Furthermore, these results add support to the hypothesis that in nanostructures the phonon are confined and have no interaction with the external environment. Unfortunately, the reflectivity of a single nanostructure could not be measured because of the difficulty in focusing the infrared light on such a small region; however the measurements of the confined phonon on a single nanostructure is being pursued in our laboratory using near-field optical techniques.

4. Conclusion

Phonon scattering by surface structuring is illustrated in the case of SiC. Infrared reflectivity measurements have been converted into real and imaginary parts of the material dielectric function. The results have been fitted to Lorentz model that considers coupled oscillators and frequency-dependent damping coefficient. We have found that the surface structuring enhances the phonon scattering mechanisms up to an order of magnitude depending on the frequency range. We hope that these results stimulate further experiments on tailoring the harmonicity and anharmonicity of the phonons for controlling the SiC infrared properties for the use in SiC-based devices. More analysis taking into account the localized phonon modes and the effect of surface structuring on the decay channels for optical phonons will be the subject of another paper.

5. Acknowledgements
M.K. and A. A.M. acknowledge the Lebanese National Centre for Scientific Research (LCNRS) for supporting this work.

References
[1] F. Wang, X. Qin, S. Ren, L. Yang, Y. Meng, and Y. Ming. Preparation and photoluminescence of netlike and curved ß-sic nanostructures. *Materials Research Bulletin*, 48(9):3640 – 3643, 2013.

[2] M. Filipescu, F. Stokker-Cheregi, D. Colceag, A. Nedelcea, R. Birjega, L.C. Nistor, and M. Dinescu. Morphological and structural characterization of sic based composite nanostructures. *Applied Surface Science*, 278(0):96 – 100, 2013. Laser materials processing for micro and nano applications, E-MRS 2012 Symposium V.

[3] E. Ziambaras and P. Hyldgaard. Thermal transport in sic nanostructures. *Materials Science and Engineering: C*, 25(58):635 – 640, 2005. Current Trends in Nanoscience - from Materials to Applications Proceedings of the European Materials Research Society 2004 - Symposium G Current Trends in Nanoscience - from Materials to Applications Proceedings of the European Materials Research Society 2004 - Symposium G Current Trends in Nanoscience - from Materials to Applications Proceedings of the European Materials Research Society 2004 - Symposium G.

[4] J.Y. Fan, X.L. Wu, and Paul K. Chu. Low-dimensional sic nanostructures: Fabrication, luminescence, and electrical properties. *Progress in Materials Science*, 51(8):983 – 1031, 2006.

[5] M. Kazan and S. Volz. Calculation of the lattice thermal conductivity in granular crystals. *Journal of Applied Physics*, 115(7), 2014.

[6] M. Shamseddine, M. Kazan, and M. Tabbal. Model for the unpolarized infrared reflectivity from uniaxial polar materials: Effects of anisotropy, free carriers, and defects. *Infrared Physics and Technology*, 55(1):112 – 121, 2012.

[7] N. Rahbany, M. Kazan, M. Tabbal, R. Tauk, J. Jabbour, J. Brault, B. Damilano, and J. Massies. Measurement of the effect of plasmon gas oscillation on the dielectric properties of p- and n-doped $Al_xGa_{1-x}N$ films using infrared spectroscopy. *Journal of Applied Physics*, 114(5), 2013.

IOP Conf. Series: Materials Science and Engineering **68** (2014) 012002 doi:10.1088/1757-899X/68/1/012002

Evidence for a large phononic band gap leading to slow hot carrier thermalisation

S Chung*[1], S Shrestha[1], X Wen[1], Y Feng[1], N Gupta[1], H Xia[1], P Yu[2], J Tang[2], G Conibeer[1]

[1]Australian Centre for Advanced Photovoltaics, School of Photovoltaics and Renewable Energy Engineering, UNSW Australia, Sydney, Australia
[2]Research Center for Applied Sciences, Academia Sinica, Taipei, Taiwan

E-mail: *simon.chung@student.unsw.edu.au

Abstract. It has been proposed that the rate of hot carrier thermalisation can be slowed down if there is a sufficiently large gap in the phonon dispersion for a bulk material. This phenomenon is critical for the development of high efficiency hot carrier solar cells to minimise energy loss to thermalisation. A gap where the minimum of the optical branches is at least twice that of the maximum of the acoustic branches can prevent the primary pathway where optical phonons loses energy, the Klemens' decay mechanism. The large gap in the phonon dispersion eliminates the Klemens' decay pathway due to energy and momentum conservation laws. This enables the electron population to remain hot by allowing sufficient time for optical phonons to re-scatter its energy to electrons. Binary compounds with a large mass difference between the two constituent atoms and high level of crystal symmetry such as zirconium nitride and hafnium nitride (HfN) have such a gap in their phonon dispersion. HfN thin films have been sputtered on silicon and quartz substrates. Characterisation of hot electron lifetimes in HfN films have been performed using ultrafast transient absorption spectroscopy. Preliminary analysis of transient absorption data, both spectra and time evolution has indicated high carrier temperatures with a nanosecond long decay time. It is postulated the long hot carrier lifetime is due to the large phononic gap.

1. Introduction

Thermalisation occurs when hot carriers relax to the conduction band edge from elevated energy levels above the conduction band. Hot carriers refer to electrons or holes with energies above the conduction band edge or below the valance band edge. The distribution of these energetic carriers can be described by the Fermi-Dirac distribution at a higher temperature, hence the name hot carriers. Thermalisation of hot carriers in photovoltaic devices makes up a significant portion of energy lost as heat which cannot be converted to electricity. The optimal semiconductor bandgap for a solar cell to achieve maximum energy conversion efficiency is characterised by a trade-off between energy lost to the lattice via hot carrier thermalisation and photons not absorbed below the bandgap [1]. At the optimal band gap of 1.3 eV as described by the Shockley-Queisser limiting efficiency of p-n junction solar cells, the energy lost due to thermalisation is 30% and below E_g loss is 25% of the incident illumination [2]. The Hot Carrier solar cell aims to address these two fundamental loss mechanisms that limit the efficiency of conventional p-n junction solar cells [3, 4, 5].

The Hot Carrier solar cell aims to capture the hot carrier population of photo-generated electron-hole pairs before thermalisation processes occurs. This enables a low band gap material to be used as a photon absorber thus minimising energy loss due to both thermalisaion and sub bandgap photons not absorbed. The hot carriers are separated and extracted to the external circuit by the use of energy selective contacts, where only carriers of an elevated energy are collected. The absorber is required to maintain a hot carrier population for a long enough time for the hot carriers to be extracted. A long hot carrier lifetime can be achieved by exploiting or engineering the phononic properties of materials.

We investigate carrier dynamics and phonon properties hafnium nitride (HfN) in the context of an absorber layer for the Hot Carrier solar cell. Materials that exhibit long hot carrier lifetimes can be used in technologies such as thermoelectric, plasmonic and thermionic emission devices.

2. Thermalisation of hot carriers and slowed carrier cooling

In a semiconductor, electrons are able to be excited into the conduction band from the valence band from absorbing the energy of the incident photon if the photon has more energy than that of the semiconductor's bandgap. Elastic carrier-carrier scattering occurs within several femtoseconds, which redistribute the energy of the hot carrier population to a Fermi-Dirac distribution described by a temperature elevated above the lattice temperature. The hot carrier population subsequently thermalises to the band edge which occurs within 10 - 100 ps. The mechanism for thermalisation involves interactions between electrons and optical phonons as described by Fröhlich interactions [6]. Energetic carriers emit optical phonons which decay into acoustic phonons. This is the point where the energy is lost and irretrievable because acoustic phonons have a large group velocity and are able to cross interfaces easily.

Decay of high energy optical phonons to lower energy acoustic phonons is dominated by processes known as Klemens decay and Ridley decay. These decay mechanisms adhere to conservation of energy and momentum laws. Klemens decay occurs when a phonon in the optical branch decays into two acoustic phonons with half the energy but opposite and equal momentum [7]. Ridley decay occurs when an optical phonon decays to an optical phonon of lower energy and an acoustic phonon [8].

It is proposed that by preventing the decay of optical phonons to acoustic phonons can slow the carrier cooling rate [9]. If there exist no pathway for optical phonons to decay into acoustic phonons a hot optical phonon population can be maintained. Electrons can reabsorb optical phonons hence maintaining a hot carrier population. A gap between the optical and acoustic branches of the phonon dispersion such that the minimum of the optical branch is at least twice that of the maximum of the acoustic branch can remove the dominate decay pathway, Klemens decay. Hafnium nitride has a wide phonon gap because of the difference in mass between the constituent atoms and the highly symmetric rock-salt crystal structure [10]. Figure 1 shows the large gap in the calculated phonon dispersion of hafnium nitride.

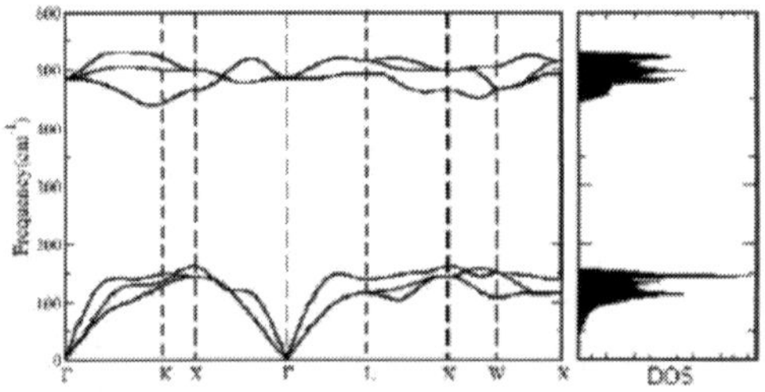

Figure 1. Phonon dispersion (left) and phonon density of states (right) of HfN [10].

3. Hafnium nitride deposition

Thin films of HfN were deposited on Si(100) and quartz substrates using an AJA ATC-2200 magnetron sputtering system by DC sputtering of a 4 inch Hf target of 99.9% purity in a mixed argon and nitrogen atmosphere. An argon gas flow rate of 15 sccm mixed with nitrogen gas flow of 1.5 sccm was used. Before deposition the deposition chamber was pumped down to a base pressure of lower than 2.5 x 10^{-6} Torr. The samples were deposited with constant DC power of 50 W and at a pressure of 1.5 mTorr. Deposition occurred with the substrates heated to approximately 450 °C. The films were deposited for 1 hour. The Si(100) substrates were chemically cleaned using RCA1, RCA2 solutions and HF dip prior to loading into the load lock chamber. Prior to the film deposition, the substrates were thermally degassed at 550 °C for 1 hr and the sputter target was sputter-cleaned for, at least 10 minutes, to remove surface oxides and other impurities.

The deposited HfN films were around 45 nm as measured using a surface profilometer. From x-ray diffraction measurements, the HfN films are polycrystalline with a preferred HfN(100) orientation on Si(100) substrates and no preferential orientation on quartz substrates. From Raman spectroscopy results, as shown in figure 2, shows close agreement between the Raman spectrum and the calculated phonon dispersion of the HfN films. The broad Raman peak around 500 cm^{-1} is associated with zone centred optical mode and Raman peak at 140 cm^{-1} is associated with the acoustic peak. Although observations of acoustic peaks are not expected via Raman spectroscopy the acoustic peak may be attributed to the non-stoichiometric HfN films leading to point defects breaking the crystal symmetry or phonon mode folding of the acoustic branches at grain boundaries of the polycrystalline film.

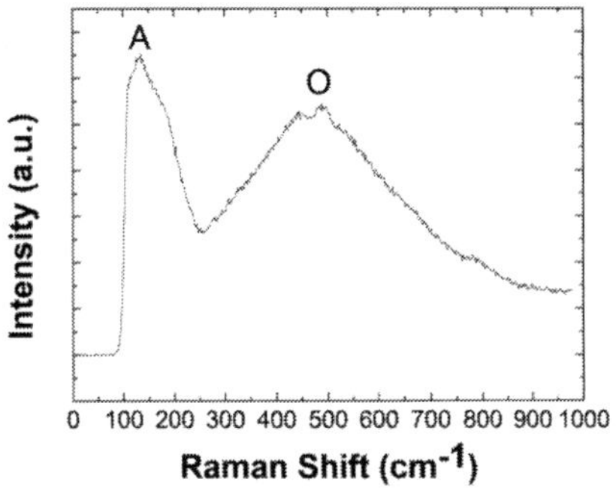

Figure 2. Measured Raman spectra of deposited HfN film.

4. Ultrafast transient absorption spectroscopy

The fs pump-probe experiments were performed with a transient absorption (TA) spectrometer (FemtoFrame II, IB Photonics). The excitation source is a 400 nm pulse by an OPA laser (TOPAS, Spectra Physics) with 100 fs duration and 1 kHz repetition rate. The probe beam of white light continuum was generated by focusing the 1 kHz Ti:Sapphire amplified laser (Spitfire, Spectra Physics) into BBO crystal, and detected by a polychromator-CCD.

Figure 3 is the TA spectrum at various time delays. Immediately after the ultra-short laser irradiation, electrons are excited to the high excited state. Ultrafast e-e scattering occurs within as short as 400 fs, a bleaching peak appears in 500-700 nm and an excited state absorption band at 470 nm. This suggests the HfN contains many free electrons and has effective e-e scattering. After this short period, a stable (ns lifetime) linear-like distribution forms in the TA spectrum,

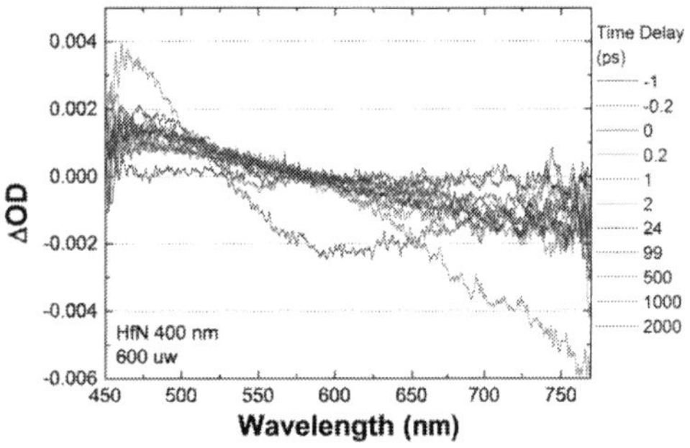

Figure 3. TA spectra of hfN with 400 nm excitation.

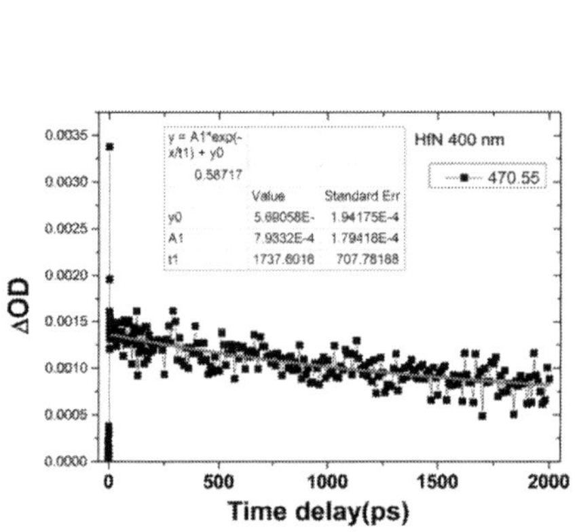

Figure 4. evolution of transient absorption at 470 nm with exponential fitting.

from 450-750 nm. This linear-like distribution takes several nanoseconds to relax back to the steady state absorption levels. Figure 4 shows the exponential fitting of the time evolution at 470 nm with exponential fitting obtaining a lifetime of 1.7 ns.

5. Conclusion

Preliminary evidence from ultrafast transient absorption spectroscopy has shown evidence of long carrier lifetimes, on the order of nanoseconds in the material hafnium nitride. It is hypothesised the long relaxation times are due to the large mass difference between the constituent atoms of HfN leading to a large gap between the optical and acoustic branches in the phonon dispersion. The gap in the phonon dispersion in HfN reduces carrier thermalisation rates by removing phonon decay pathways creating a phonon bottleneck effect allowing electrons to reabsorb optical phonons. Materials which exhibit extremely slow carrier thermalisation rates can be used for the development of the Hot Carrier solar cell which can theoretically achieve very high power conversion efficiencies.

Acknowledgements

This work has been supported by the Australian Government through the Australian Renewable Energy Agency (ARENA). Responsibility for the views, information or advice expressed herein is not accepted by the Australian Government.

References

[1] Shockley W and Queisser H J, 1961 *Journal of Applied Physics* **32**(3) 510
[2] Hirst L C and Ekins-daukes N J, 2011 *Progress in Photovoltaics: Research and Applications* **19**(3) 286
[3] Ross R and Nozik A, 1982 *Journal of Applied Physics* **53**(5) 3813
[4] Würfel P, Brown A S, Humphrey T E and Green M A, 2005 *Progress in Photovoltaics: Research and Applications* **13**(4) 277
[5] Green M A, 2003 *Third Generation Photovoltaics : Advanced Solar Energy Conversion.* Springer-Verlag Berlin and Heidelberg GmbH & Co. KG. ISBN 9783540265627
[6] Fröhlich H, 1954 *Advances in Physics* **3**(11) 325
[7] Klemens P, 1966 *Physical Review* **148**(2) 845
[8] Ridley B, 1996 *Journal of Physics: Condensed Matter* **8** L511
[9] Conibeer G, König D, Green M and Guillemoles J, 2008 *Thin Solid Films* **516**(20) 6948
[10] Saha B, Acharya J, Sands T D and Waghmare U V, 2010 *Journal of Applied Physics* **107**(3) 033715

EMRS Spring 2014 D IOP Publishing
IOP Conf. Series: Materials Science and Engineering **68** (2014) 012003 doi:10.1088/1757-899X/68/1/012003

Cavity modes and optomechanic interactions in strip waveguide

Said El-Jallal[1,2], Mourad Oudich[1], Yan Pennec[1], Bahram Djafari-Rouhani[1], Abdelkader Makhoute[2], J. Gomis-Bresco[3], D. Navarro-Urrios[3], A. Martínez[4] and C. M. Sotomayor Torres[3, 5]

[1]*Institut d'Électronique, de Microélectronique et de Nanotechnologie, Université de Lille 1, Villeneuve d'Ascq, France.* [2]*Physique du rayonnement et de l'interaction Laser-matière, Université Moulay Ismail, Faculté des sciences, Meknès, Maroc.* [3]*ICN2 - Institut Catala de Nanociencia i Nanotecnologia, Campus UAB, 08193 Bellaterra (Barcelona), Spain* [4]*Nanophotonics Technology Center, Universitat Politècnica de València, Valencia, Spain* [5]*ICREA - Institucio Catalana de Recerca i Estudis Avançats, 08010 Barcelona, Spain*

E-mail: said.el-jallal@ed.univ-lille1.fr

Abstract: Phoxonic crystals can exhibit dual phononic/photonic band gaps. Therefore, the confinement of both acoustic and optical waves in a phoxonic cavity can allow the enhancement of their interaction. In this paper, we discuss our recent theoretical works on the strength of the optomechanic coupling, based on both photoelastic and moving interfaces mechanisms, in nanobeam phoxonic crystals cavities.

1. Introduction

There has been a great deal of recent interest about optomechanical interaction in different structures based on slabs[1-3] and nanobeams[4-5] using well-defined cavities. However, most of these works considered localized phonons and photons that are not necessarily trapped inside a photonic and/or a phononic band gap. In this paper, we study optomechanic interactions in phoxonic crystals which are defined as dual phononic/photonic crystals that can exhibit simultaneously phononic and photonic band gaps. The existence of absolute band gaps allows the simultaneous confinement of both waves that, in turn, can produce the enhancement of their interaction for the purpose of novel and high-performance optomechanical and acousto-optic devices and applications. A main objective is the modulation of light by acoustic waves when both excitations are confined inside the same cavity or propagate with a slow velocity inside a waveguide. We have studied theoretically this optomechanic interaction in different (2D[6,7], slabs[8], and strip[11,12]) phoxonic crystals cavities. This paper will focus on the results of the strip waveguide structures.

2. Mechanisms and Methods

Two mechanisms contribute to the acousto-optic interaction, namely the photoelastic (PE) and moving interface (MI) effects[6,7]. The former is due to a local variation of the dielectric permittivity induced by the acoustic strain inside the materials (Pockels effect[9]), whereas the second mechanism comes from the variation of the dielectric permittivity in the vicinity of the interfaces due to the motions of the boundaries during the vibrations of the cavity. To evaluate the strength of the phonon-photon coupling, we calculate the modulation of the cavity photon frequency by the cavity phonon, namely the photonic mode frequency is calculated at several selected instants of an acoustic period under the assumption that the acoustic mode strain profile is being frozen at these instants[6,8]. This is justified by the fact that the photon frequency is higher by several orders of magnitude than the acoustic frequency. The acousto-optic strength is proportional to the amplitude of this photonic frequency modulation. The calculation should be done for each pair of phononic and photonic modes in order to compare the strength of the coupling between different pairs. The coupling may vanish to the first order for some combinations of phononic and photonic modes due to symmetry reason. From our calculation, we can evaluate the relative contributions of the PE and MI effects and, in particular, see whether both effects are in phase and add to each other or they are out of phase and partly cancel each other. Finally, the above analysis can be compared to the optomechanical coupling coefficient[5] that expresses the photonic frequency shift induced by the zero-point motion of the mechanical field of the phonon and which can be calculated from the knowledge of the acoustic and optical field distributions inside the cavity.

3. Strip waveguide

We have investigated phoxonic crystals constituted by 1D strip (or nanobeam) waveguides. The corrugated nanobeam is constituted by a backbone periodically drilled with holes and along which lateral stubs are periodically attached[10]. We have shown that the holes and the stubs are favorable for the opening of the photonic and phononic gaps, respectively, so this structure is suitable to provide simultaneous phononic and photonic gaps. Then, we have studied the possibility of cavity modes as well as slow guided modes in these structures[10].
In our previous work[10], we studied a simple cavity by introducing a silicon space defect between two holes. This cavity displayed both confined acoustic and optical modes. However, the quality factor of the photonic mode inside the gap was very low (less than 1000) which makes this cavity not attractive for the experimental observation of the optomechanic coupling. To overcome this disadvantage, we investigate here the design of new cavities providing both high photonic quality factors and strong optomechanic coupling coefficients.

The realization of a high quality factor can be achieved by tailoring a tapered cavity in which the geometrical parameters of the unit cells change progressively from the cavity center towards both sides to reach their values in the perfect crystal. For instance, by using a parabolic variation of the parameters when the parameters at the cavity center are about 70% of those in the perfect crystal, we are able to achieve quality factors as high as 2 to 3 10^5 at wavelengths of about 1500nm.

Before evaluating the strength of the optomechanic coupling, it is worth noticing that, due to symmetry reasons, only acoustic modes having a symmetric-symmetric (SS) behavior with respect to the two symmetry planes of the nanobeam can couple to photonic modes. In addition, if the cavity itself has a symmetry plane at its middle, the considered phonon should also be symmetric with respect to this third plane and can be called SSS. Using the tapered cavity structure described above, we calculated the strength of the optomechanic coupling between all the photonic cavity modes and all the SSS acoustic modes up to 5 GHz (see Fig. 1). The phononic modes falling in the range of the absolute band gap (3.5 to 4 GHz) give relatively weak acousto-optic coupling, up to g~0.006 MHz. The strongest coupling of g~2.3 MHz (with PE and MI contributions of respectively 1.6 and 0.7 MHz) is obtained with an acoustic mode at 2.46 GHz which is not inside a gap and thus corresponds to a resonant, rather than to a localized, mode of the cavity. However, this mode is very close to a phononic gap of SS symmetry and below the full band gap, so we can expect that by changing slightly the shape of the tapered cavity, it can be pushed into the partial gap or above inside the full band gap without changing significantly the optical modes of the cavity. The modification of the cavity consists of increasing or decreasing the length of the stubs progressively from the perfect crystal towards the cavity center. The coupling factor is significantly increased when the acoustic mode is pushed inside the full band gap. This modification does not change significantly the coupling factor, but the mechanical quality factor of the acoustic mode pushed inside the partial band gap should be much improved. This can make the structure more attractive for the experimental observation of the optomechanical interaction. Figure 1 illustrates the results as concerns the geometry of the two kinds of strip and the corresponding coupling coefficients g between some photonic and phononic modes.

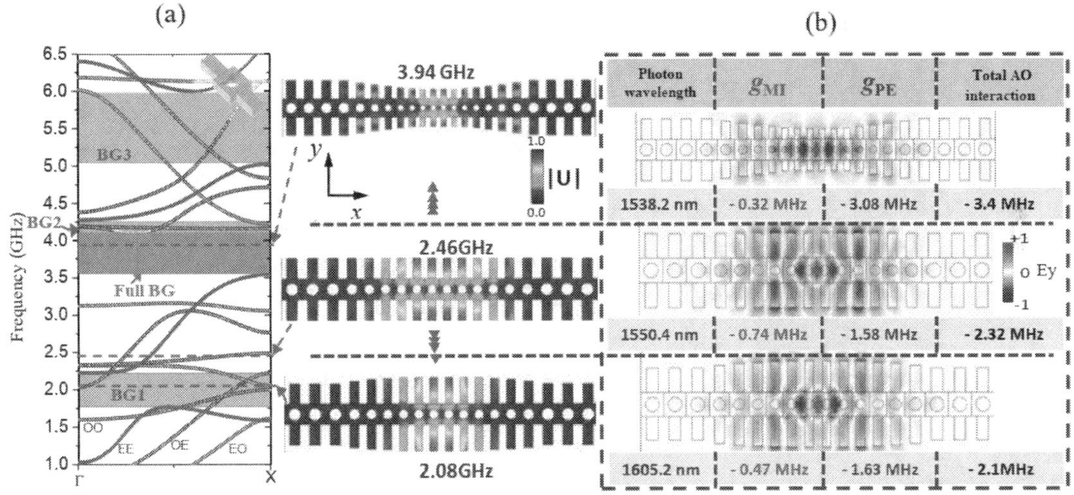

Figure1: (a) Phononic band structure of the perfect nanobeam where the red shaded regions named BG1, BG2 and BG3 are the gaps for the modes which are symmetrical with respect to the symmetry planes of the nanobeam (red branches) and which are the only one that are able to couple with photonic modes and the gray shaded region for the full band gap. (b) AO coupling between the photonic cavity modes and the same phononic mode pushed up inside the full band gap or down inside the partial band gap by decreasing or increasing respectively the length of the stubs. In each figure we give the coupling coefficients g_{PE} and g_{MI}.

We present an example of strip waveguide where phononic and photonic cavity modes give strong AO coupling. The total coupling coefficient is evaluated at 2.1 MHz where the MI and PE effects contribute additively (see figure 2).

4. Conclusion

In conclusion, we have investigated the AO interaction in a corrugated phoxonic silicon nanobeam presenting band gaps both for photonic and phononic modes. By creating a well-designed tapered cavity, we were able to create highly confined phononic and photonic modes with high quality factor inside the gap which can then enhance their AO interaction. We have shown that this tapering can be adjusted to shift either the photon wavelength or the phonon frequency of interest in the suitable targeted wavelength or frequency range which can be useful for experimental AO phenomena observation and characterization. In particular, high mechanical quality factor can be achieved when the phonon mode that couples efficiently to photons is pushed inside a gap by choosing an appropriate tapering of the heights of the stubs.

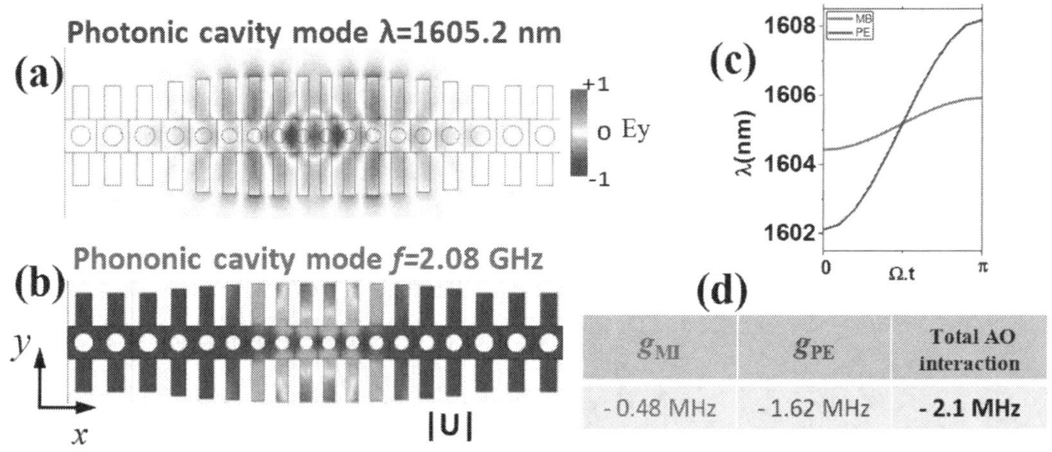

g_{MI}	g_{PE}	Total AO interaction
- 0.48 MHz	- 1.62 MHz	**- 2.1 MHz**

Figure 2. (a) Electric field for the photonic mode, (b) norm of displacement field for the phononic mode, (c) modulation of the photonic mode by the acoustic cavity mode during half acoustic period, (d) the coupling coefficients g_{PE} and g_{MB} for photoelastic (blue solid line) and moving boundaries (red solid line) effects respectively.

5. References

[1] A. H. Safavi-Naeini, T. P.Mayer Alegre,M.Winger, and O. Painter, Appl. Phys. Lett. **97**, 181106 (2010).

[2] D. A. Fuhrmann, S. M. Thon, H. Kim, D. Bouwmeester, P. M. Petroff, A. Wixforth, and H. J. Krenner, Nature Photon. **5**, 605 (2011).

[3] E. Gavartin, R. Braive, I. Sagnes, O. Arcizet, A. Beveratos, T. J. Kippenberg, and I. Robert-Philip, Phys. Rev. Lett. **106**, 203902 (2011).

[4] M. Eichenfield, J. Chan, R. M. Camacho, K. J. Vahala, and O. Painter, Nature **462**, 78 (2009).

[5] J. Chan, A. H. Safavi-Naeini, J. T. Hill, S. Meenehan, and O. Painter, Appl. Phys. Lett. 101, 081115(2012).

[6] Y. Pennec, B. Djafari-Rouhani, C. Li, J. M. Escalante, A. Martinez, S. Benchabane, V. Laude, and N. Papanikolaou, AIP Adv. **1**, 041901 (2011).

[7] Q. Rolland, M. Oudich, S. El-Jallal, S. Dupont, Y. Pennec, J. Gazalet, J. C. Kastelik, G. Leveque, and B. Djafari-Rouhani "Acousto-optic couplings in two-dimensional phoxonic crystal cavities" Applied Physics Letters **101**, 061109 (2012).

[8] S. El-Jallal, M. Oudich, Y. Pennec, B. Djafari-Rouhani, V. Laude, J.C. Beugnot, A. Martinez, J. M. Escalante, and A. Makhoute "Analysis of optomechanical coupling in two-dimensional square lattice phoxonic crystal slab cavities" Phys. Rev. B 88, 205410 (2013).

[9] D. Royer and E. Dieulesaint, Elastic Waves in Solids II: Generation, Acousto-optic Interaction, Application, Springer, (1999).

[10] Y. Pennec, B. Djafari-Rouhani, C. Li, J. M. Escalante, A. Martinez, S. Benchabane, V. Laude, and N. Papanikolaou, AIP Adv. **1**, 041901 (2011).

[11] M. Oudich, S. El-Jallal, Y. Pennec, B. D. Rouhani, J. Gomis-Bresco, D. Navarro-Urrios, C. M. S. Torres, A. Martinez, and A. Makhoute, "Optomechanic interaction in a corrugated phoxonic nanobeam cavity", Phys. Rev. B 89, 245122 (2014).

[12] J. Gomis-Bresco, D. Navarro-Urrios, M. Oudich, S. El-Jallal, A. Griol, D. Puerto, E. Chavez, Y. Pennec, B. D. Rouhani, F. Alzina, A. Martinez and C.M. Sotomayor Torres, "A one-dimensional optomechanical crystal with a complete phononic band gap", nature communications 5, 4452 (2014).

IOP Conf. Series: Materials Science and Engineering **68** (2014) 012004 doi:10.1088/1757-899X/68/1/012004

Conversion of Laser Pulse Optical Energy to Photo-acoustic Wave in nm-Scale Layered $TlGaSe_2$ Crystals

K Gulbinas[1], V Grivickas[1], P Grivickas[2], J Linnros[3]

[1] Institute of Applied Research, Vilnius University, Vilnius, Lithuania
[2] Institute for Shock Physics, Department of Physics and Astronomy, Washington State University, Pullman, Washington, USA
[3] School of ICT, Royal Institute of Technology, Kista-Stockholm, Sweden

E-mail: Karolis.Gulbinas@ff.stud.vu.lt

Abstract. Experiments are presented that reveal an efficient optical energy conversion from the visible to the infrared wavelengths range as a result of photo-acoustic response (PAR) after light pulse incites onto the free surface of $TlGaSe_2$ crystal. Excitation was carried out with a tunable wavelength of ns-pulse laser and the PAR was detected laterally with a focused cw-probe. The observed properties can be related to variety of successive factors: high electron-hole-phonon deformation potential, a high factor of refraction coefficient dependency on pressure, the absence of surface recombination and the band filling effect, in relation with low absorption coefficient due to the forbidden direct-band optical transition in $TlGaSe_2$. All these ensure that the acoustic energy remain well confined under a wide pulse power and energy range suggesting that $TlGaSe_2$ is a promising material for dynamic optical energy conversion.

1. Introduction

The ternary compound $TlGaSe_2$ is a dense quasi-two dimensional layered semiconductor with revealed predominantly forbidden electron transitions and a close indirect and direct optical band gaps at $E_G \approx$ 2.1 eV resulted from a weak covalent nature of the interlayer bonding [1]. The layer consists of corner linked $GaSe_4$ tetrahedra where Tl atoms lay on straight lines along *a,b* crystalographic **c**-axis of trigonal voids between tetrahedra. Layer stacking direction is along the growth **c**-axis and every consequent layer is twisted by 90° in stacking plain. Within the layer prevails strong covalent bonding while between the layers interaction is weak. The structure undergoes sequential phase transitions below RT. Because of this and some other extraordinary properties, $TlGaSe_2$ crystals received a great attention for applications in optoelectronics [1-3]. We present prominent acoustic responses in $TlGaSe_2$ which can be used for pulsed energy conversion in optical circuits. The results are obtained by the method of the time-resolved acoustic wave perpendicular probing. It is based on probe deflection under the perpendicular probe direction to the longitudinal wave propagation on a rectangular semiconductor solid strip. The technique was adopted from acoustic studies in liquids and gasses and is referred as a schlieren optical probe method [4]. The preliminary results have been published in [5].

2. Experimental

The $TlGaSe_2$ samples used for PAR measurements were grown by modified Bridgman method from high purity elements taken in stoichiometric proportions. Optical quality surfaces parallel to the layers (001) were cleaved mechanically and no additional surface treatment was needed. Perpendicular

IOP Conf. Series: Materials Science and Engineering **68** (2014) 012004 doi:10.1088/1757-899X/68/1/012004

lateral surfaces however are not smooth; the edges are crumbled and have a step like layered nature. It must be noted, that the preparation of samples with specular lateral faces was a difficult task. A careful gentle polishing thus was performed along the *a (*or *b)* axis which are a natural termination for layer edges [6]. This allowed to form optical quality lateral facets. Rectangular corners were achieved by additional removal of some damaged part by cleavage.

The sample was glued with silver paste to the holder allowing (001) facet excitation along $\|c$ direction, while probe beam was directed perpendicularly $\perp c$ through the lateral surface, as shown in figure 1. Tuneable pulse laser (*Infinity Coherent*) used for excitation was of 2 ns pulse duration and of 40 Hz repetition rate. Excitation beam was focused to $d = 0.9$ mm spot. The narrow probe-beam (from a typical infrared LED) focused to $d \sim 7$ μm penetrated lateral surface. It could be positioned at any distance from the excited surface with 1 μm precision [5]. After probe was transmitted throughout the sample, it was collimated with two lenses. A knife edge was inserted between the lenses to block a portion of the beam. This allows detecting a small variation of the refractive index generated by acoustic wave; see more quantitative description of physical principles below. Next, the transmitted probe beam was focused to 0.5 ns rise time InGaAs photo- receiver and the signal trace was observed on 2 GHz oscilloscope and recorded to PC. Typically the 500 traces were averaged to increase the signal-to-noise ratio. The measurements were performed in the vacuum cryostat at room temperature (RT) and at 77 K.

After excitation pulse hits the free surface of the absorbing sample, a fraction of absorbed light energy goes to the generation of acoustic wave. In agreement with earlier studies, we expected completely decoupled mechanisms contributions of electronic stress ($d_{CV}\Delta N$, where d_{CV} is a plasma-phonon deformation potential and ΔN is concentration of the excited electrons and holes) and of thermoelastic stress induced in the media. The thermoelastic stress is released as a thermal energy of hot carriers relaxation energy (E - E_G) and by a delayed nonradiative recombination in the plasma [7]. Since the carrier and the heat diffusivity in $TlGaSe_2$ across the layers are very small we expected identical stress profiles to contribute as a simple sum under our experimental conditions. Assuming that the total stress only depend on depth z, the longitudinal acoustic (LA) strain profile travelling into the sample by sound velocity v_s is expressed by $\varepsilon(\eta) \propto (1/\rho V_s^2)[3B\beta\Delta T(0,t) + d_{CV}\Delta N]/F(\eta)$, where $\eta = z - v_3 t$, V_s is thermal capacity volume, B is the bulk modulus, β is the linear expansion coefficient and $F(\eta)$ is a waveform function which has been analytically given by bipolar pressure waveforms with the duration of double optical absorption length ($1/\alpha$) [8]. The equation predicts that the initial strain polarity on a free surface depends on the sign of d_{CV} and gives rise to the contraction or dilatation stresses accordingly. The predicted pressure wave form is shown in the middle inset of the figure 1.

The method of a schlieren optical probe provides time-resolved acoustic wave probing based on the probe deflection mode under the perpendicular probe transition to the LA wave propagation [4]. The

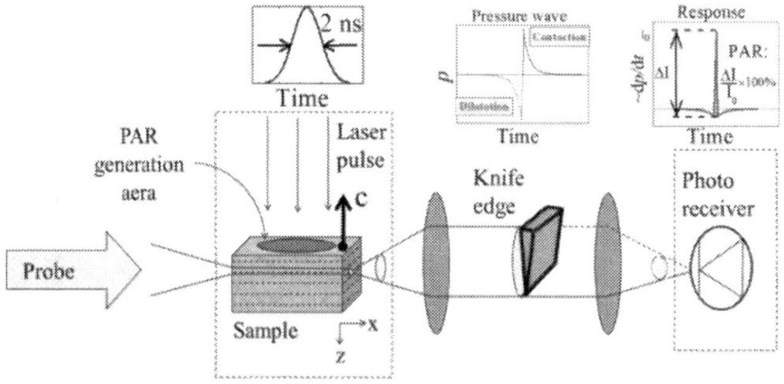

Figure 1. A scheme for excitation direction $\|$c in layered $TlGaSe_2$. Pump pulse excites the surface area and the narrow probe beam penetrates the sample perpendicularly to the longitudinal acoustic wave propagation direction. The insets explain waveforms of the acoustic stress and the PAR response.

schlieren scheme (figure 1) is sensitive due to relatively wide intersection area (~1 mm) between the LA wave and the probe. Also this method has no shear acoustic components. The acoustic stress dilatation and contraction cause the deflection of opposite directions providing the temporal derivative of the refractive index versus strain pressure profile in the transverse plane. The transfer function of refractive index variation *vs.* pressure [4] can be considered as a simple linear function since the acoustic stress of the LA amplitude is on a milibar level [9]. Important to note that if one would block an opposite side of the probe beam with the knife edge one would get an opposite event registration succession, i.e. the obtained PAR would be of the opposite polarity [5]. In the result section we have adopted PAR as a ratio between the signal amplitude and the ground level as shown in the right inset of figure 1. Majority of performed measurements were obtained with 1540 nm probe, while with 633, 671, 861, 1300 nm probe beams quite similar results arise in $TlGaSe_2$.

3. Results and discussion

Figure 2 shows PAR echoes in $TLGaSe_2$ for 6.3 mJ/cm^2 excitation fluence at different quanta at RT in a 300 μm-thick sample. PAR shapes are of symmetrical form and qualitatively are explained by considering that dilatation and contraction are proportional to deflection angle, as was predicted (see figure 1). We determine that the acoustic wave propagates along the **c**-axis with longitudinal speed of sound $\upsilon_{ac} = 2.63 \cdot 10^5$ cm/s [5,6]. When acoustic wave reaches back side of the sample it reflects and experiences phase shift by π which interchanges the contraction-dilatation order of the stress, therefore, PAR propagating backwards appears of the same polarity in the transient (see 2nd PAR instance in figure 2a). The third (3) and fourth (4) PAR are reflections from the front and from the back surfaces, respectively. This process continues for several microseconds (figure 2b) with low dispersion of acoustic energy [4]. That indicates high optical quality (small roughness) of the cleaved surfaces. Under a low excitation quanta (< 2.0 eV) the absorption coefficient is low enough to allow excitation through the bulk of the sample. Then, the PAR signal superimposes on the background shape of free-carrier-absorption. Also, in this spectral range, we observe small PAR generated from the back surface propagating in opposite direction and having opposite polarity. (The PAR depicted in figure 2a as (1`) - generated from the back surface and (2`) is a reflection from the front surface.) This behavior was explicitly investigated in reference [5] showing that, at the specific probe depth positions, PAR signals from the front surface and from the back surface is superimposed and can be virtually canceled.

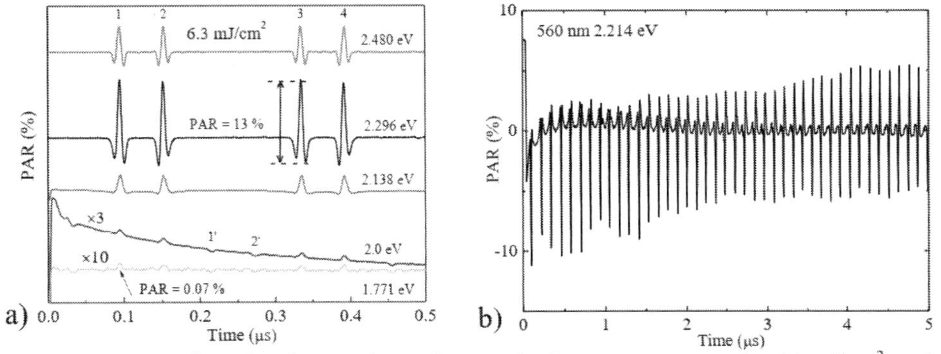

Figure 2. (a) PAR transients in $TlGaSe_2$ for various excitation quanta energy for 6.3 mJ/cm^2 excitation in 300 μm-thick sample. An indexed instant of echoes, namely, 1st is PAR generated on front surface; 2nd – reflection from the back; 3rd – reflection from front surface; 4th – second reflection from front surface. The reflected PAR changes stress polarity, therefore is registered as the same dilatation process. Note that at low energies excitation penetrates whole sample and PAR generation from back surface arises simultaneously which appear of the opposite polarity: 1` - generated; 2'- reflected. Also free carrier absorption signal is pronounced for 2.0 eV curve. (b) Multiple echoes of PAR maintain constant form in μs-time scale indicating a good optical quality of the cleaved opposite surfaces.

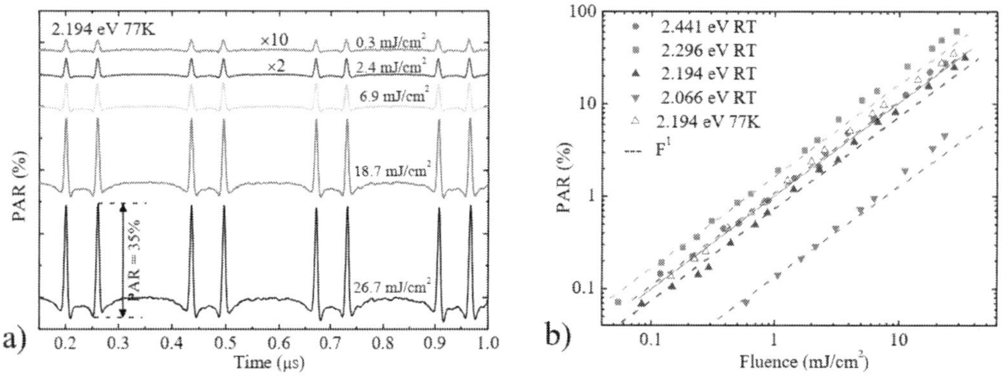

Figure 3. (a) PAR transients for different excitation fluencies at 2.194 eV and at 77 K temperature. (b) PAR increase with increasing laser fluence at different quanta energy at RT and at 77 K. In a wide spectral range PAR show linear growth $\sim F^1$ as given by fitted lines.

Figure 3a shows PAR measured at different excitation fluence of 2.194 eV at 77 K. PAR growth is observed over three orders of excitation laser fluence F. Experiments were carried out in a wide spectral range 1.75 – 2.7 eV. Wave form remains the same shape though some FWHM narrowing occurs from 18 ns to 8 ns due to absorption depth shrinkage with increasing absorption coefficient approaching 2.7 eV. The growth follows almost a linear dependence ($\sim F^1$) at all spectral range. Few transients on depicted fluencies are shown in figure 3b. The small sublinear add-on we explain by an increase of the thermoelastic stress contribution due to emerging of a fast recombination lifetime component of an order of the laser pulse duration. This component was depicted form free-carrier absorption transient (not shown). The PAR magnitudes of 70 % were reached with no sign of PAR saturation. Moreover, PAR wave form remains undistorted at high excitations at all times. These two facts emerge from predominantly forbidden electron optical transitions and a close indirect and direct optical band gaps resulted from a weak nature of the interlayer bonding in $TlGaSe_2$ [6].

In figure 4a it is shown PAR dependence versus absorption coefficient α (figure 4b) as measured by free-carrier absorption at 77 K [5]. For excitation energy above the indirect exciton edge (2.08 eV) PAR linearly increases with α until the absorption depth reaches acoustic energy delivery depth $1/\alpha =$ $= \tau_L \times \upsilon_{ac}$. This is predicted by theory of acoustic response [4] and is called as a condition of proper stress confinement. When absorption depth becomes shorter than the acoustic delivery depth, the PAR begins to decrease. For $1/\alpha \ll \tau_L \times \upsilon_s$, the theory predicts inverse dependence (PAR $\sim 1/\alpha$) [4]. Both the linear and inverse dependences are shown in figure 4a by fitted lines and the acoustic delivery point at 1.9×10^3 cm^{-1} is indicated by an arrow. Agreement with theory is quite satisfactory for the above the band gap excitation, i.e. at $\alpha > 40$ - 3000 cm^{-1} (figure 4b). However, higher PAR is observed than the theory predicted for the below band gap excitation at $\alpha < 40$ cm^{-1} (figure 4a). This finding we explain as deviation from the bipolar condition, i.e. the electron concentration is low compared to hole concentration, and as possibility of excess hole dislocation on a different layers on a ns-scale layer near the surface of the crystal [6]. Lose od locality between excess hole and trapped electron can cause additional contribution to PAR similar to the piezoelectric mechanism observed on GaAs (111) surfaces [7].

The deformation potential d_{CV} and refraction coefficient dependence on pressure (*dn/dp*) still are not established by the direct experiments in $TlGaSe_2$. Combination of these factors contribute to high PAR magnitudes and determine the PAR polarity. PAR polarity in $TlGaSe_2$ is certainly opposite than in GaAs [10]. We assume that *dn/dp* in $TlGaSe_2$ is positive as in layered InSe and GaSe layered crystals. Assuming the same value for $TlGaSe_2$ ($dn/dp = 0.045$) we estimate $d_{CV} \cong 3$ eV in comparison with known value for GaAs $d_{CV} = 10$ eV. Hence high PAR is caused mainly by the large *dn/dp* value.

The absence of surface recombination, restricted carrier diffusivity across the layers and forbidden optical transitions are the main factors determining linear PAR dependence on carrier injection.

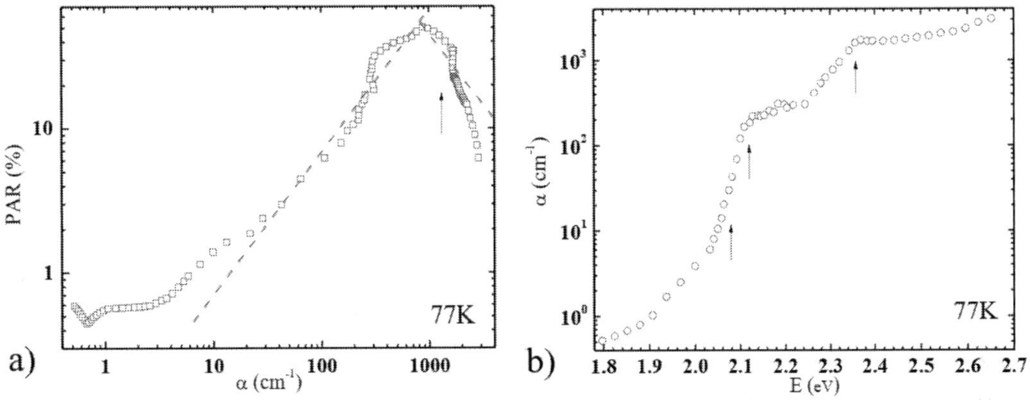

Figure 4. (a) PAR as a function of absorption coefficient at 77 K and (b) absorption spectrum obtained by free-carrier absorption [5]. The characteristic indirect exciton at 2.08 eV and two direct ones at 2.12 eV and at 2.36 eV are indicated.

4. Conclusions

We have applied a schlieren optical probe method for the detection of LA waveforms in $TlGaSe_2$ plates under perpendicular probe direction to crystallographic **c**-axis. PAR exhibits symmetric strain pulse generated by coinciding shapes of electronic and thermoelastic stresses because of low carrier diffusion and heat transfer across the crystal layer plain. Such LA pulses propagate millimeter distances in the sample under μs-time range with low acoustic dispersion. PAR is observed from the subband gap excitation spectral range from 1.7 eV (at $\alpha \sim 1$ cm^{-1}) into the above band gap range until 2.7 eV (at $\alpha > 5 \times 10^3$ cm^{-1}). The PAR echoes remain of the same waveform shape with a linear dependence versus excitation energy up to high value of PAR = 70% for appropriate optical excitation fluence ~ 30 mJ/cm^2. The latter property gives rise to the very specific features of indirect-direct band gap nature with mainly forbidden optical transitions. Large PAR can be used for optical light pulse conversion into the infrared wavelength signals and easy acoustical signals integration into an optical circuits.

Acknowledgements
This study was supported by Swedish Visby program project No. 00729.

References

[1] Panich A M 2008 *J. Phys.: Condens. Matter*. **20** 293202
[2] Seyidov M Y, Suleymanov R A and Salehli F 2012 *Phys. Lett.* A **376** 643
[3] Johnsen S, Liu Z, Peters J A, Song J-H, Peter S C, Malliakas C D, Cho N K, Jin H, Freeman A J, Wessels B W and Kanatzidis M G 2001 *Chem. Mater.* **23** 3120
[4] Sigrist M W 1986 *J. Appl. Phys.* **60** R83
[5] Grivickas V, Bikbajevas V, Gavryushin and Linnros J 2007 *J. Phys.: Conf. Ser.* **100** 042007
[6] Grivickas V, Gulbinas K, Gavryushin V, Bikbajevas V, Korolik O V, Mazanik A V and Fedotov A K 2014 *Phys. Status Solidi* RRL (unpublished)
[7] Young E S K, Akimov A V, Campion R P, Kent A J and Gusev V 2012 *Phys. Rev.* B **86** 155207
[8] Thomsen C, Grahn H, Maris H and Tauc J 1986 *Phys. Rev.* B **34** 4129
[9] Song K, Cha H, Lee J and Veselovskii I A 1995 *Appl. Phys.* B **61** 547
[10] Grivickas V, Gulbinas K and Grivickas P 2012 in HPSP-15, Montpelje, France, Abstract P2.11 (unpublished)

Sensitive 3-omega measurements on epitaxial thermoelectric thin films

Y Q Liu[1], D Tainoff[1], M Boukhari[2], J Richard[1], A Barski[2], P Bayle-Guillemaud[2], E Hadji[2] and O Bourgeois[1]

[1] University Grenoble Alpes, CNRS, Institut Néel, Grenoble F-38042, France
[2] Institut Nanosciences et Cryogénie, SP2M, CEA-UJF, 17 Rue des Martyrs, Grenoble 38054, France

E-mail address: yanqing.liu@neel.cnrs.fr

Abstract. Here we present a work on the development of a sensitive 3-omega instrumentation adapted for the measurement of the thermal conductivity of epitaxial thin films. The experimental setup has been validated by the measurements on an epitaxial germanium nanostructured thin film. This manganese doped germanium matrix contains Ge_3Mn_5 nanoinclusions having a diameter of 5 to 50 nm, grown by Molecular Beam Epitaxy (MBE). The 3 omega measurements have revealed that the thermal conductivity of GeMn nanostructured thin films can be decreased by a factor of ten as compared to the bulk value.

1. Introduction

With the development of the nanotechnologies, the domain of thermoelectricity has pushed itself to the frontiers of the global research on energy harvesting. For the characterization of the ZT value (figure of merit) of a thermoelectric material, experimental challenges lie in a precise measurement of the thermal conductivity. Efforts are aimed at finding efficient thermoelectric materials under the idea of "electron crystal-phonon glass" material system [1]. The efficiency of a thermoelectric material is characterized by the dimensionless figure of merit (*ZT* value), which involves the Seebeck coefficient S, electrical conductivity σ and thermal conductivity κ through the formula of:

$$ZT = \frac{S^2 \sigma T}{\kappa}. \tag{1}$$

However, highly accurate measurements of the three concerned factors are required for a straightforward determination of the *ZT* value. The measurement of thermal conductivity stays the greatest experimental challenge, especially for epitaxial thermal conductive thin films grown on electrical conductive substrates, which are good candidates for potential thermoelectric applications in microelectronics.

The two most developed measurement techniques for thermal conductivity are the 3-omega method [2] and the time-domain thermoreflectance (TDTR) [3]. The TDTR is a pump-probe optical technique, where the time evolution of the surface temperature is measured through temperature-dependent changes in the reflectivity (i.e., the thermoreflectance) [4, 5]. Though the TDTR gives a good spatial resolution, the 3-omega method still shows its advantage in the accuracy of the measurement on thermal resistive thin films [6, 7].

In the work here, we would like to present an advanced experimental instrumentation based on the 3-omega method, developed for highly precise measurements of thermal conductivity of epitaxial semi-conductor thin films. The performance of the technique has been confirmed by measurements on a germanium based epitaxial thin film, being a highly promising thermoelectric material. The 3-omega measurements of this thin film contribute to the investigation of the thermoelectric properties of the material, also help us to achieve qualitatively and quantitatively a better understanding of the thermal transport mechanism in nanostructured materials.

2. Experimental methods

2.1. Sample preparation

To investigate the thermal properties of a thin film using the 3-omega method, a transducer metal line acting both as a heater and thermometer is needed to be deposited on the film. In our case it is designed to have a geometry as shown in figure 1. The length of the transducer between the two internal contacts is 2 mm, with its width being 30 μm. The electric current is driven through the two external contacts and the voltage is obtained via the two internal ones. Standard clean room processes like UV lithography and PVD are used to deposit a platinum transducer. Platinum is a common material used for thermometry, it has a temperature coefficient of $\alpha \sim 0.003\ K^{-1}$. The thickness of the Pt deposition is 100 nm, so that the transducer gives a resistance of around 100 Ω at RT in formerly described geometry.

For the case of an electrical conductive thin film, an oxide layer is required to electrically isolate the transducer from the sample surface. The aluminium oxide (Al_2O_3) is chosen to do the task (figure 2) in our work. Using the technique of ALD (Atomic Layer Deposition) the thickness of Al_2O_3 thin film is well controlled to be 50 nm, which has been tested previously to sustain a current up to 10 mA.

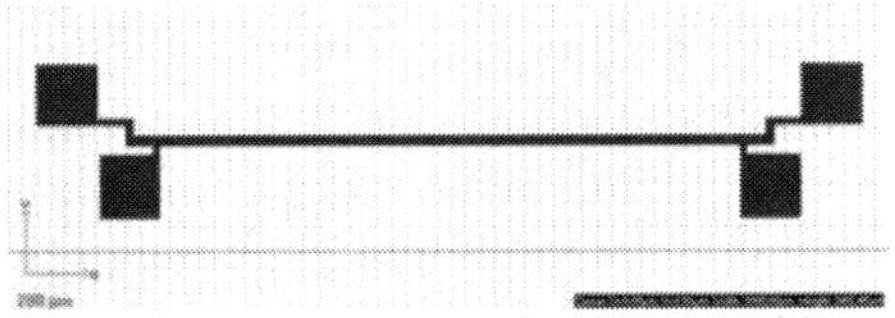

Figure 1. Designed motif in lithographic mask for the deposition of transducer.

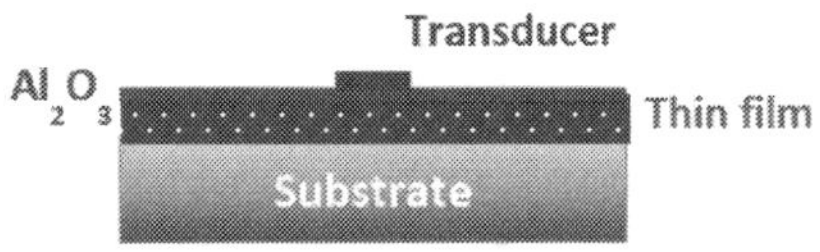

Figure 2. Schematic illustration of the sectional structure of a prepared sample for thermal measurement.

2.2. Setups for 3-omega measurement

The 3-omega method is about measuring the third harmonic part ($V_{3\omega}$) of the voltage signal detected at the two extremities of the transducer when applying an alternating current (AC) of angular frequency ω. The frequency dependence of this $V_{3\omega}$ permits the extraction of the thermal conductivity of the substrate material.

Since the $V_{3\omega}$ voltage is 1000 times smaller than the first harmonic voltage $V_{1\omega}$, it cannot be measured without the help of an active electronic setup increasing the $V_{3\omega}/V_{1\omega}$ ratio. Two solutions are used in most of the cases. One is based on electronic techniques in a geometry of "differential bridge" [2, 3, 6], and another one by using the structure of Wheatstone bridge [8, 9]. Here in our case, we have designed our own device based on the idea of a differential bridge; the overall electrical circuit of the setup is shown in figure 3. The variable resistance of the device R_{ref} is highly precise to 0.001 Ω, and highly stable with temperature.

IOP Conf. Series: Materials Science and Engineering **68** (2014) 012005 doi:10.1088/1757-899X/68/1/012005

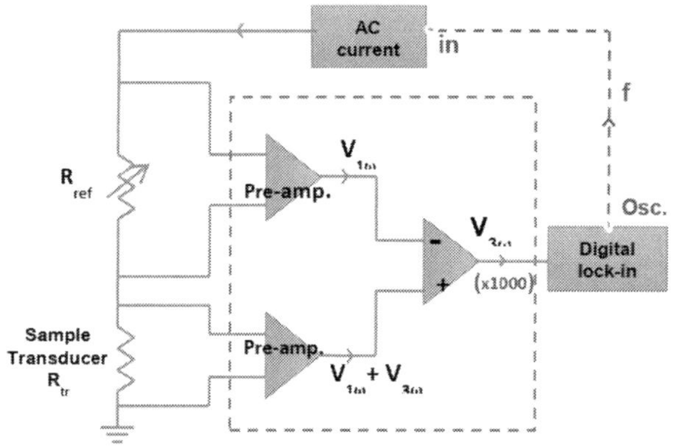

Figure 3. Schematic diagram of the electrical circuit (differential bridge) for the 3-omega measurement (the area inside the dashed line represents a home-made electronic device).

The sample is pasted by a resist to the sample holder of a calorimeter, after the micro-bonding, the holder is mounted inside a cryostat, closed, pumped and connected, ready for the measurement. The cryostat ensures a large range of temperature environment from the temperature of liquid helium up to 400K.

3. Modelling of 3-omega method for thin films and strategy of measurement

3.1. The 3-omega method for thin film

According to the principles of 3-omega method, for the case of a narrow transducer metal line on the surface of an infinite half-volume substrate, an approximation of the temperature oscillation at the sample surface can be deduced [2] if the width of the line is largely small comparing to the thermal penetration depth (low frequency range), and the thickness of the transducer is negligible. The realisation of the first condition strongly depends on the frequency range used for measurement. As a general trend, the frequency range is distinguished by two regimes, the high frequency regime and the low one, by comparing the width ($2b$) of the transducer metal line to the thermal penetration depth (μ) into the related sample substrate. The thermal penetration depth is defined by:

$$\mu = \left(\frac{D}{\omega}\right)^{-1} \qquad (2)$$

where D is the thermal diffusivity of the substrate material below the transducer. For the need of the approximation of the surface temperature oscillation, the frequency regimes are defined by:

Low frequency regime: $\mu >> 2b$ (3)

High frequency regime: $\mu << 2b$

The second condition can easily be satisfied because usual substrates are generally much thicker (~ 500 μm) than the transducer line (~ 100 nm). In the low frequency regime the temperature oscillation at the sample surface can thus be approximated by :

$$\Delta T = \frac{P_l}{\pi \cdot k_0}\left(-\frac{1}{2}\ln\frac{\omega}{\Omega} + \eta - i\frac{\pi}{4}\right) \qquad \left(P_l = \frac{R_{tr} I^2}{l}\right) \qquad (4)$$

where P_l is the linear heat dissipation with R_{tr} the electric resistance, I the applied AC current and l the length of the transducer; the thermal angular frequency $\omega = 4\pi f$, where f is the electric frequency of the AC current; $\Omega = \frac{k_0}{\rho C b^2}$ with k_0 the thermal conductivity of the substrate, ρ the density, C the specific heat, b the half width of the transducer and $\eta = 0.92$.

Via the relation between the real part of the temperature oscillation and the real part of the $V_{3\omega}$ signal shown in equation (4), the thermal conductivity of the sample substrate k_0 can be calculated through the equation:

$$V_{3\omega nph.} = \frac{1}{2} R_{tr} I \alpha \cdot \Delta T_{real} \tag{5}$$

$$k_0 = -\frac{\alpha \cdot R_{tr}^2 \cdot I^3}{4 \cdot \pi \cdot l}\left(\frac{dV_{3\omega inph.}}{d\ln\omega}\right)^{-1} = -\frac{1}{2 \cdot \pi \cdot l}\left[\frac{d\left(\frac{\Delta T_{real}}{P}\right)}{d\ln\omega}\right]^{-1} \tag{6}$$

where $\frac{d\left(\frac{\Delta T_{real}}{P}\right)}{d\ln\omega}$ can be obtained through a linear fit of the curve $\frac{\Delta T_{real}}{P}(\ln\omega)$ within the concerned frequency zone

Our thin film of interest (thickness t_1) is generally deposited on a very thick substrate (Si or Ge or other). As illustrated in figure 4, as long as t_1 is largely small comparing to the width $2b$ of the transducer, the heat transport is considered to be one-dimensional across the film and the heat flux conserved.

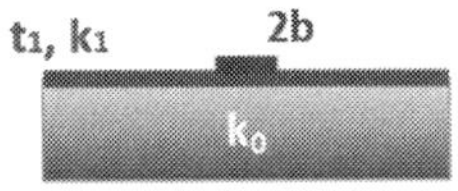

Figure 4. Schematic illustration of a sample substrate with a thin film deposited on top (in red) that needs to be measured.

This approximation brings a shift in the real part of the surface temperature oscillation:

$$\Delta T = \frac{P_l}{\pi \cdot k_0}\left(-\frac{1}{2}\ln\frac{\omega}{\Omega} + \eta - i\frac{\pi}{4}\right) + \frac{P_l \cdot R'}{2b} \tag{7}$$

Or

$$\frac{\Delta T_{real}}{P} = \frac{1}{\pi \cdot l \cdot k_0}\left(-\frac{1}{2}\ln\frac{\omega}{\Omega} + \eta\right) + \frac{R'}{2bl} \tag{8}$$

$$\Delta\left\langle\frac{\Delta T_{real}}{P}\right\rangle = \frac{R'}{2bl} \tag{9}$$

where the thin film represents itself only as a thermal barrier resistance R' for the heat flux, which is directly related with its thickness and thermal conductivity through:

$$R' = \frac{t_1}{k_1} + R_c \tag{10}$$

where R_c is the effective thermal resistance of interfaces, being the total interface thermal resistance of the thin film involved sample minus the total one for the case of the sample substrate.

3.2. Measurement strategy

Following the above model, measurements need to be carried out on two sets of samples to extract the thermal conductivity of an electrical conductive thin film, as illustrated in figure 5. It is the ΔT shift between them that will give the thermal conductivity of the layer of interest.

As a test of the method, the thermal conductivity of an Al_2O_3 thin film is investigated using the model and the experimental modus operandi explained above. Measurements are carried out on samples of an Al_2O_3 thin film (50 nm) on top of a GaAs substrate (electrical resistive), and of the GaAs substrate alone as a reference sample. The thermal conductivity of the Al_2O_3 thin film is found to be around 0.8 $Wm^{-1}K^{-1}$ (considering R_c to have a value of 10^{-8} Km^2W^{-1} as an approximation [10]), showing a nice coherence with published relative works [11].

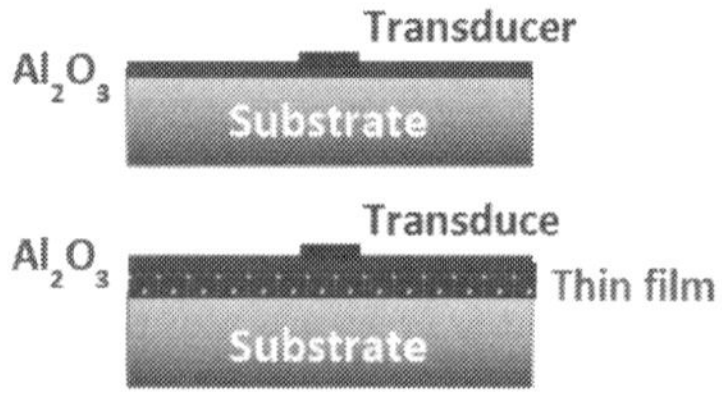

Figure 5. Schematic illustration of the sample geometries for the 3-omega measurement of an electrical conductive thin film.

Further tests have been done on an epitaxial thin film, based on a well crystallized and highly doped germanium matrix, and containing spherical Ge_3Mn_5 nano-inclusions [12, 13]. The nano-inclusions are randomly distributed in the matrix having a diameter of 5 to 50 nm depending on the epitaxial conditions, e.g. the concentration of Mn, growing temperature, annealing temperature and etc. These Ge:Mn thin films are grown on germanium substrate with a thickness of 100 to 300 nm in the general cases.

According to the sample geometries and the structural properties of Ge:Mn thin films, some approximations are considered for the determination of the effective thermal resistance of interfaces for the sample pair. As the thin film of Ge:Mn is grown by molecular beam epitaxy on the Ge substrate using MBE, the germanium matrix of Ge:Mn thin film is well crystallized as its Ge substrate, thus the interface thermal resistance between Al_2O_3 and Ge is considered to be equal to the one between Al_2O_3 and Ge:Mn:

$$R_{Al_2O_3/Ge} \cong R_{Al_2O_3/Ge:Mn} \tag{11}$$

The effective interface thermal resistance R_c can then calculated via:

$$R_c = (R_{Pt/Al_2O_3} + R_{Al_2O_3/Ge:Mn} + R_{Ge:Mn/Ge}) - (R_{Pt/Al_2O_3} + R_{Al_2O_3/Ge}) = R_{Ge:Mn/Ge} \tag{12}$$

and the interface thermal resistance between the epitaxial Ge:Mn and Ge ($R_{Ge:Mn/Ge}$) is relatively very small down to the order of 10^{-9} Km^2W^{-1} [14, 15]. The R_c can then be determined to be:

$$R_c \approx 10^{-9} (K \cdot m^2 \cdot W^{-1}) \tag{13}$$

4. Experimental results

The 3-omega measurements have been carried out on n-type Ge substrate and Ge:Mn thin films grown on the same n-type doped Ge substrate. The thin film is 240 nm thick, with the concentration of manganese 6%. As for our samples concerning a germanium substrate (D = 3.6x10^{-5} m^{2}s^{-1}) being around 0.35 mm thick, based on the calculation of μ as shown in table 1, the frequency range for our 3-omega measurements is chosen to be between 100 Hz and 1000 Hz. Data of $V_{3\omega}$ signals as a function of frequency from the sample and the reference Ge (n-type) substrate, at different temperature, are gathered and treated. The curves of $\frac{\Delta T_{real}}{P}$ as a function of $\ln\omega$ for cases at different temperatures are plotted in figure 6.

Table 1. Thermal penetration depth into the Ge substrate at frequency 100 Hz and 1000Hz.

f **(Hz)**	100	1000
μ **(µm)**	169	54

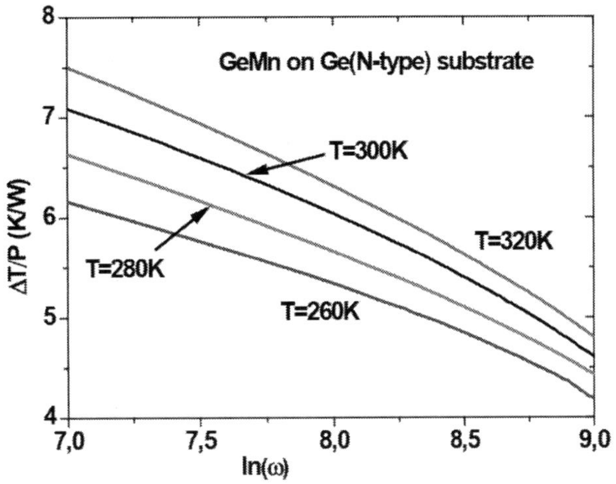

Figure 6. Signals of $\frac{\Delta T_{real}}{P}$ as a function of $\ln\omega$ at four different temperatures (260K, 280K, 300K and 320K) measured from Ge:Mn film on Ge substrate.

With the slope from a linear fit of the curves, the conductivity of the Ge substrate (below the thin films in the Ge:Mn sample) can be calculated. The same treatment has also been done for signals from reference substrate sample, and the results for the bulk thermal conductivity from both samples are presented in table 2. The differences between these two set of measurement is a good indication on the absolute error (accuracy) made on the measurement of k. This error in accuracy can be estimated to be around 5%.

Table 2. Results of the thermal conductivity of Ge (n-type) substrate as a function of temperature for the two different sets of sample: bare Ge and Ge/Ge:Mn.

T (K)		260	280	300	320
k_{Ge} ($W.m^{-1}.K^{-1}$)	Ge substrate	74.7	64.8	58.4	56.7
	Ge:Mn sample	81.2	70.1	60.6	57.7

At a given temperature, the comparison of the curves $\frac{\Delta T_{real}}{P}$ from the two samples reveals a clear shift as shown in figure 7, from the curve of reference sample to the Ge:Mn sample. At 320K in the figure, based on the shift of a value of 0.75, the thermal conductivity of the Ge:Mn thin film is calculated to be 5.5 (±1) $Wm^{-1}K^{-1}$. This remarkable reduction of the thermal conductivity as compared to the bulk Ge substrate is currently attributed to the nanostructuring occurring in the GeMn layer. Extensive measurements are in progress to fully understand the influence of this nanostructuring on the thermal properties of GeMn thin films.

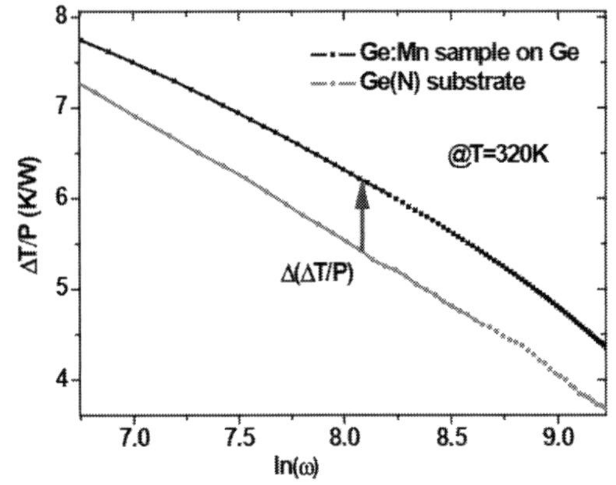

Figure 7. Comparison of the curves $\frac{\Delta T_{real}}{P}$ from Ge:Mn sample and the Ge reference substrate sample at 320K.

5. Conclusions

Based on the 3-omega method, we have developed a sensitive instrumentation for the characterisation of thermal properties of thin films. The measurements of the thermal conductivity have been successfully performed on an epitaxial nanostructured germanium thin film. The results have revealed a remarkably reduced thermal conductivity of this thin film by a factor of 10 at 320K, compared to the value for bulk Ge substrate (60W/mK).

To properly understand the physics of the involved phonon scattering mechanism introduced by the nanostructures, the 3-omega measurements will be continued on different samples of the Ge:Mn thin film with further investigations into the influence of the nano-inclusions (size and spacial distribution) on the thermal property: i.e. the Mn concentration, the diameter and dispersion of the inclusion. Further experiments will also be carried out in parallel for a complete characterisation of the thermoelectric properties of the thin film, including the measurements of the electrical conductivity, the Seebeck coefficient, and finally the determination of the ZT value.

Acknowledgements
We acknowledge the financial support from the European project MERGING grant agreement N° 309150, and from the Région Rhône-Alpes through the project COGEETEN of the ARC Energy program. We also thank G. Julié, E. André and J.-L. Garden as well as Nanofab and Captercal technological pools for their valuable technical support.

References

[1] Slack G A 1995 *Handbook of Thermoelectrics* ed D M Rowe (Boca Raton) p 407-40
[2] Cahill D G 1990 *Rev. Sci. Instrum.* **61** 802
[3] Cahill D G *et al.*, Applied Physics Reviews **1** 011305
[4] Cahill D G 2004 *Rev. Sci. Instrum.* **75** 5119
[5] Huxtable S, Cahill D G, Fauconnier V, White J O and Zhao J C 2004 *Nat. Mater.* **3** 298
[6] Cahill D G, Katiyar M and Abelson J R 1994 *Phys. Rev.* B **50** 6077
[7] Borca-Tasciuc T, Kumar A R and Chen G 2001 *Rev. Sci. Instrum.* **72** 2139
[8] Wojciechowski K T, Zybala R and Mania R 2009 *J. Achi. Mater. Manufact. Eng.* **37** 512-7
[9] Sikora A, Ftouni H, Richard J, Hébert C, Eon D, Omnès F and Bourgeois O 2012 *Rev. Sci. Instrum.* **83** 054902, and Sikora A, Ftouni H, Richard J, Hébert C, Eon D, Omnès F and Bourgeois O 2013 *Rev. Sci. Instrum.* **84,** 029901
[10] Saci A 2011 *Transport thermique dans les milieux nano-structurés (GaAs)n / (AlAs)n* (Paris : Ph.D. thesis)
[11] Cappella A, Battaglia J L, Schick V, Kusiak A, Lamperti A, Wiemer C and Hay B 2013 *Adv. Eng. Mater.* **15** 1046
[12] Devillers T 2008 *Ferromagnetic phases of Ge(1-x)Mn(x) for spintronics applications* (Grenoble: Ph.D. thesis)
[13] Jain A, Jamet M, Barski A, Devillers T, Yu I S, Porret C, Bayle-Guillemaud P, Favre-Nicolin V, Gambarelli S, Maurel V, Desfonds G, Jacquot J F and Tardif S 2011 *J. Appl. Phys.* **109** 013911
[14] Lee S M, Cahill D G and Venkatasubramanian R 1997 *Appl. Phys. Lett.* **70** 2957
[15] Costescu R M, Wall M A and Cahill D G 2003 *Phys. Rev.* B **67** 054302

IOP Conf. Series: Materials Science and Engineering **68** (2014) 012006 doi:10.1088/1757-899X/68/1/012006

Acousto-optic couplings in two-dimensional Lithium Niobate phoXonic crystal

Quentin ROLLAND[1], Samuel DUPONT[1], Joseph GAZALET[1], Jean-Claude KASTELIK[1]

[1]Institut d'Electronique, de Micro-électronique et de Nanotechnologie, UMR CNRS 8520, Université de Valenciennes-Hainault-Cambrésis, Valenciennes, France

*Corresponding author: quentin.rolland@univ-valenciennes.fr

Abstract. We investigate the acousto-optic couplings in two-dimensional infinite piezoelectric phoXonic crystal cavities. The periodic structure considered consists in a square array of air inclusions drilled in Lithium Niobate matrix. This structure allows the simultaneous confinement of acoustic and optic waves in a point defect obtained by removing one hole from the structure. Our theoretical analysis takes into account three different coupling mechanisms. In addition to the traditional effects: photo-elastic effect and opto-mechanic effect existing in non-piezoelectric materials, we consider the electro-optic effect introduced by the electric field accompanying the phonon. Using the finite element method, we analyze the perturbation of the photonic cavity modes under the action of phononic cavity modes. We first discuss some aspects of the additional electro-optic effect. For illustration purposes, we compare the coupling strength of two analogous breathing modes, one as found in the LiNbO3 structure and another mode qualitatively comparable but evidenced in a similar silicon structure. Then, we end the discussion by an analysis of the coupling strength dependence on typical crystallographic cuts of Lithium Niobate.

1. INTRODUCTION

PhoXonic crystal designates a nanostructure that exhibits simultaneously photonic [1,2] and phononic [3-5] crystals behaviors. The capital letter X stands for both the characters t and n which refer to photonic and phononic. The strong interest for phoXonic devices is mainly triggered by the fact that these structures lead to acousto-optic devices with very promising efficiencies [6]. Indeed, these manmade crystals provide a tool for confining simultaneously photonic and phononic energy, in relation with slow-waves and/or forbidden frequency bandgaps phenomena that affect both light and sound waves. Beside the expected high efficient acousto-optic interaction, devices based on phoXonic crystals reduce the energy consummation owing to their miniaturization as compared to conventional acousto-optic devices [7]. Furthermore, phoXonic crystals offer the possibility to control simultaneously light and sound in frequency domains of interest for optical communication systems. For instance, the technology advances allow the design of PhoXonic crystals with periods extending a few hundreds of nanometers. At these dimensions, the optical operating wavelengths fall in the telecommunication ranges while the acoustic frequencies are in the GHz range.

Maldovan and Thomas [8] have demonstrated the existence of simultaneous photonic and phononic forbidden bandgaps in 2D square lattice realized by air holes drilled in bulk silicon. Other authors have extended the study to other materials and patterns. For example, simultaneous bandgaps have been studied in sapphire [9], in lithium niobate [10] for different lattices like honeycomb or triangular lattices. More recently, researches on simultaneous bandgaps have been focused on 3D calculation including slabs [11, 12], pillars [13] or strip waveguides [14]. Also, acousto-optic modulators based on isolated cavities [6-15] and waveguides [16] have been investigated in silicon phoXonic crystal. Recently [17] we have demonstrated that square and honeycomb lattices, correctly designed on silicon and lithium niobate, may develop sufficiently wide optical and acoustic forbidden bands to provide simultaneous confinement of optical and acoustic modes. However, triangular lattice presents too thin acoustic bandgaps to ensure this simultaneous confinement.

In the present letter we study the acousto-optic modulation through the photoelastic, the optomechanic and the electro-optic effects in silicon and lithium niobate in different situations. This enables to get insights on the influence of the media on the acousto-optic couplings: by a judicious choice of some couples of mode (photonic-Phononic) we analyze the various interaction mechanisms and give an interpretation of their relative importance.

2. THEORETICAL MODEL AND MATHEMATICAL PROCEDURE

In a first step, the crystal phononic and photonic band structures are determined using the conventional unit-cell method combined with the Bloch-Floquet boundary conditions. Then, the localized cavity modes are established using a 7x7 supercell model. At last, the acousto-optic modulation is characterized in a quasi-static approach. This is fully justified considering the 5 orders of magnitude separating the phononic (Ω) and the photonic (ω) frequencies. In other words, the localized acoustic modes (instantaneous mechanical displacement vectors, deformation tensors and the companion piezoelectric fields in the case of LiNbO3 material) are determined at discrete time instants within an acoustic period $0 \leq \Omega \cdot t < 2\pi$. The resulting instantaneous maps of the acoustically disturbed permittivity are reintroduced within a multi-physics procedure, in order to determine the optical perturbed modes. The acousto-optic modulation results are presented as the photonic angular eigenfrequency (ω) excursions during a complete acoustic period $0 \leq \Omega \cdot t < 2\pi$.

From a phenomenological point of view, one can distinguish different effects contributing to the perturbation of the permittivity by the acoustic modes. Namely, the photo-elastic i.e. Pockel effect, and the opto-mechanical effect resulting from the moving hole boundaries. A third effect must also be taken into account in case of piezoelectric materials like $LiNb0_3$. This is the electro-optic effect initiated by the piezoelectric field accompanying the deformation field.

The 2D model is developed using a commercial Finite Element Method (FEM) software. Concerning the photonic crystals modeling, all domains are meshed. Whereas, only solid materials domains are meshed for phononic crystal simulations, since elastic waves cannot propagate in air. For the computation, we use the values of the stiffness tensor, the mass density and the refractive index as published in [17] and [18] for silicon and lithium niobate respectively. Finally, the photo-elastic and electro-optic parameters are taken from [19] and [20] for silicon and lithium niobate respectively.

3. RESULTS AND DISCUSSION

All the presented results are given for square lattices with period a, of air holes included in bulk materials. The hole reduced radius $r/a = 0.48$ is chosen to provide bandgaps for both waves in silicon and lithium niobate simultaneously. The acoustical eigenmodes are labeled according to the ascending eigenfrequency order using capital letters in silicon and with primed capital letters in lithium niobate. Similarly, the optical eigenmodes are labeled with the same convention but with Greek letters. In both materials the eigenmodes are characterized by their eigenfrequencies and eigenvectors: the field distributions.

In order to compare the acousto-optic coupling mechanisms in the different materials, we have chosen acoustic modes with similar displacement spatial distributions: mode F in silicon and mode C' in LiNbO3, as illustrated in fig.1 inset. Both modes are "breathing" modes: these particular modes induce a change of the cavity volume through a contraction and dilatation without too many distortions. The color scale depicts the total displacement amplitude whereas the arrows indicate the displacement direction. Optical TE modes α and β' have been chosen for the same reason: modal similarity among the two materials. They are both characterized by an E_z field distribution with two principal lobes. For the TM polarization, no optical modes with such a degree of similarity have been found. So, we have made the choice of optical modes δ and ε in silicon and modes α' and β' for lithium niobate: both of these modes share the particularity of presenting four H_z lobes. We present here the most representative results of an exhaustive numerical study to assess the effect of the different coupling mechanisms on the overall optical mode modulation.

Figure 1.a presents the modulation of optical TE modes α and β' eigenfrequencies under the perturbation of acoustic modes F and C' respectively as a function of the acoustic period $\Omega \cdot t$ varying between 0 and 2π. As can be seen, in Silicon the modulation consecutive to the elasto-optic effect is not significant and the total modulation effect is due to the optomechanic effect. The results in Lithium Niobate are quite different: first one can observe that both optomechanic and elasto-optic effects produce significant modulations and are out of phase (consecutively their total contribution is less significant than one effect considered alone), whereas the electro-optic contribution is not significant. Such observations can be explained qualitatively by an inspection of the physical constant tensor elements at stake in the model. Strictly speaking, the electro-optic contribution would depend on r_{33} electro-optic coefficient. But, in the case of a strictly 2D model considered here (i.e. the system is assumed to be invariant with respect to translation in the Z axis direction) this dependence no longer holds. On the other hand, the photoelastic constants at stake are: $p_{12} = 0.017$ in Silicon and $p_{31} = 0.17$ in Lithium Niobate. The factor ten between the constants is in correlation with the modulations observed.

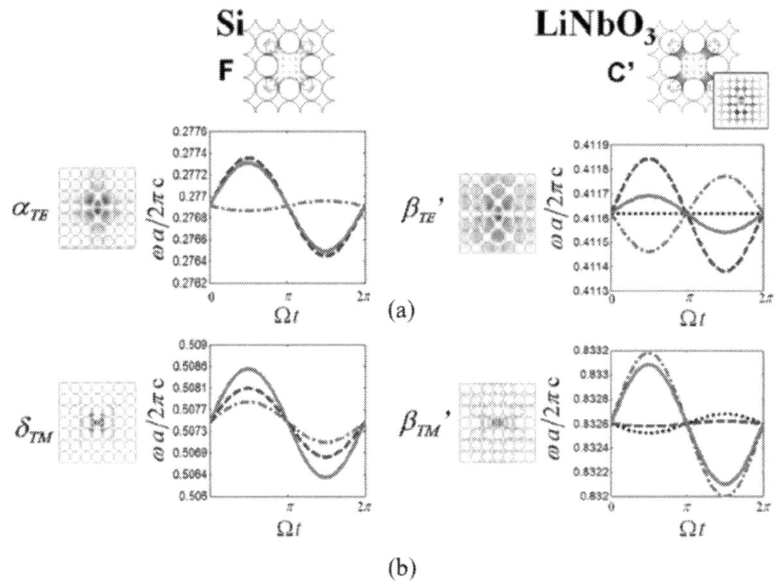

Figure 1: Modulation of optical eigenmode frequencies as a function of one acoustic period, in silicon structure (left) and $LiNbO_3$ Z-cut (right). Fig 1.a (top) modulation of TE modes, fig. 1.b (bottom) modulation of TM modes. The green, blue and black curves correspond to the modulation due to the photo-elastic, the optomechanic and the electro-optic effects respectively. The red curve presents the overall modulation when all acousto-optic mechanisms are taken into account.

Figure 1.b presents the modulation of the optical TM modes α and β' eigenfrequencies consecutive to an acoustic perturbation respectively by modes F and C', as a function of the acoustic period: $0 \leq \Omega \cdot t < 2\pi$. As it can be seen in this case, the modulation induced by the elasto-optic effect is quite similar in both materials. The photo-elastic tensor elements that are involved in the coupling now are a combination of: $p_{11} = 0.09$ and $p_{44} = 0.05$ in Silicon and $p_{12} = 0.09$ and $p_{11} = 0.026$ in Lithium Niobate.

The photoelastic constants at stake are quite similar in both materials, hence a modulation with the same order of magnitude is found.

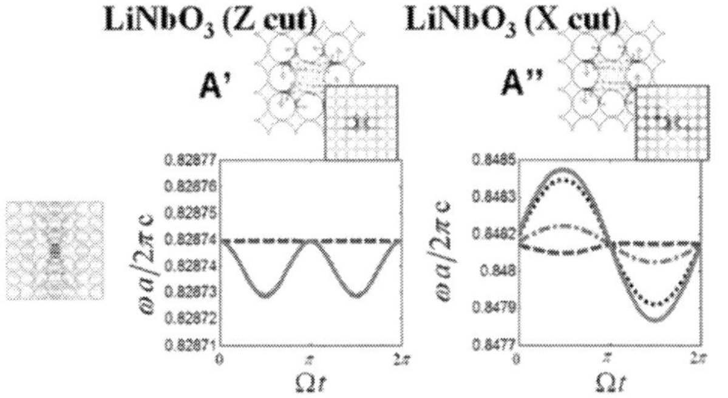

Figure 2: Modulation of optical eigenmode frequencies as a function of one acoustic period, in $LiNbO_3$ Z-cut structure (left) and $LiNbO_3$ X-cut (right). The green, blue and black curves correspond to the modulation due to the photo-elastic, the optomechanic and the electro-optic effects respectively. The red curve presents the overall modulation when all acousto-optic mechanisms are taken into account.

Figure 2 now compares two situations for the same material LiNbO3, but for different crystallographic cuts. The modulation of the optical TM modes α' eigenfrequencies consecutive to an acoustic perturbation by modes A' in Z-cut and X-cut respectively is displayed as a function of the acoustic period $\Omega \cdot t$. At first sight, it is surprising that two acoustic modes sharing approximately the same displacement distribution and modulating two optical modes with very similar field distributions present quite significant different acousto-optic modulation profiles: those of the Z-cut are characteristic of a second order effect whereas the X-cut ones are typical of a first order effect. The second observation concerns the electro-optic effect. Practically unnoticeable in the case of a Z-cut material, it becomes the strongest effect in the case of an X-cut one. The latter observation is easily linked to the electro-optic coefficient at stake: for Z-cut, $r_{22} = 3.4\ 10^{-12}$ V/m is at stake in the coupling whereas $r_{13} = 7.7\ 10^{-12}$ V/m, $r_{22} = 3.4\ 10^{-12}$ V/m, $r_{42} = 18\ 10^{-12}$ V/m and $r_{33} = 28.8\ 10^{-12}$ V/m are all together involved in the electro-optic coupling in X-cut. Concerning the second / first order behavior of the acousto-optic perturbation, we remind that the perturbation theory tells us that the first order correction vanishes whenever the perturbation owns an anti-symmetric plane and so, the usually weak second order effects are no more hidden and manifest themselves. This is exactly what happens in the case of the Z-cut LiNbO3. But, in the X-cut case, while at a first glance the acoustical displacement field seems to be very similar to that of the Z-cut, if the distribution of the perturbed permittivity tensor elements is carefully inspected, it reveals that the antisymmetric plane does not exist anymore. We attribute this observation to the strong acoustic anisotropy in the case of X-cut. [15,17]

4. CONCLUSION

Considering two materials, silicon and lithium niobate, we have established that the acousto-optic coupling between an optical and an acoustical cavity mode is strongly influenced by the choice of the couple of modes. The different interaction mechanisms can sustain or oppose each other. We give an interpretation of the modulation behavior based on an analysis of the materials intrinsic parameters. As it is established for propagating waves, different interaction configurations result in different interaction efficiencies, this is also the case here for cavity modes. On the other hand, the strong anisotropy of lithium niobate influences the acoustic field distribution: different cuts provide different modes. Moreover, when quasi-similar modes are found on different cuts, they may lose symmetry elements which drastically modify the optical modulation. The crystallographic cut is found to be a crucial factor in the design process of phoXonic crystals for acousto-optic interaction.

References

[1] E. Yablonovitch, *Phys. Rev. Lett.* **58**, 2059 (1987)
[2] J. D. Joannopoulos, S. G. Johnson, J. N. Winn, and R. D. Meade, Photonic Crystals: Molding the Flow of Light (*Princeton University Press*, 2008)
[3] M. S. Kushwaha, P. Halevi, L. Dobrzynski, and B. Djafari-Rouhani, *Phys. Rev. Lett.* **71**, 2022 (1993)
[4] M. M. Sigalas and E. N. Economou, *Solid State Commun.* **86**, 141 (1993)
[5] Y. Pennec, J. Vasseur, B. Djafari-Rouhani, L. Dobrzynski, and P. A. Deymier, *Surf. Sci. Rep.* **65**, 229 (2010)
[6] I. E. Psarobas, N. Papanikolaou, N. Stefanou, B. Djafari-Rouhani, B. Bonello, V. Laude, *Phys. Rev. B* **82**, 174303 (2010)
[7] N. Courjal, S. Benchabane, and J. Dahdah, G. Ulliac, Y. Gruson, V. Laude, *Appl. Phys. Lett.* **96**, 131103 (2010)
[8] M. Maldovan and E. L. Thomas, *Appl. Phys. Lett.* **88**, 251907 (2006)
[9] D. Bria, M. B. Assouar, M. Oudich, Y. Pennec, J. Vasseur, and B. Djafari-Rouhani, *J. Appl. Phys.* **109**, 014507 (2011)
[10] S. Sadat-Saleh, S. Benchabane, F. I. Baida, M. P. Bernal, and V. Laude, *J. Appl. Phys.* **106**, 074912 (2009).
[11] Y. Pennec, B. Djafari Rouhani, E. H. El Boudouti, C. Li1, Y. El Hassouani, J. O. Vasseur, N. Papanikolaou, S. Benchabane, V. Laude, and A. Martinez *Optics Express* **18**, 13 14301 (2010)
[12] Y. Pennec, B. Djafari Rouhani, E. H. El Boudouti, C. Li, Y. El Hassouani, J. O. Vasseur, N. Papanikolaou, S. Benchabane, V. Laude, and A. Martinez *Chinese Journal of Physics* **49**, 1 (2011)
[13] Y. El Hassouani, C. Li, Y. Pennec, E. H. El Boudouti, H. Larabi, A. Akjouj, O. Bou Matar, V. Laude, N. Papanikolaou, A. Martinez, and B. Djafari Rouhani, *Phys. Rev. B* **82**, 155405 (2010)
[14] Y. Pennec, B. Djafari Rouhani, C. Li,1 J. M. Escalante, A. Martinez, S.Benchabane, V. Laude, and N. Papanikolaou, *AIP Advances* **1**, 041901 (2011)
[15] Q. Rolland, M. Oudich, S. El-Jallal, S. Dupont, Y. Pennec, J. Gazalet, J. C. Kastelik, G. Lévêque, and B. Djafari-Rouhani, *Appl. Phys. Lett.* **101**, 061109 (2012)
[16] Tzy-Rong Lin, Chiang-Hsin Lin and Jin-Chen Hsu *J. of Appli. Phys.* **113**, 053508 (2013)
[17] Q. Rolland, PhD thesis « Couplages acousto-optiques dans les cristaux photoniques et phononiques » (2013)
[18] D. K, Biegelsen, *Phys. Rev. B* **32**, 21 1196 (1974)
[19] R.S. Weis and T.K. Gaylord, *Appl. Phys. A* **37** (4), 191-203 (1985)
[20] M. Jazbinsek and M. Zgonik *Appl. Phys. B* **74**, 407–414 (2002)

Theory of interface and anharmonic phonon interactions in nanocomposite materials

Iowerth O. Thomas[1] **and Gyaneshwar P. Srivastava**
School of Physics, University of Exeter, Stocker Road, Exeter, EX4 4QL, UK
E-mail: G.P.Srivastava@exeter.ac.uk; gps@excc.ex.ac.uk

Abstract.
We present a theory of phonon scattering rates resulting from mass smudging across interfaces and from anharmonicity in nanocomposite materials. We discuss the derivation of the anharmonic scattering term and present some recently published results indicating that if certain models of interface mass-mixing are adopted, the cross-planar thermal conductivity of an ultra-thin superlattice system will decrease dramatically with the number of bilayers n until $n \approx 4$, above which value it begins to increase with n.

1. Introduction

It is becoming clear that phonon properties of nanostructured materials can be 'tuned' to a large extent for future technological applications, such as thermoelectricy with an enhanced figure of merit. A key phonon property is the lattice thermal conductivity $\kappa_{\rm ph}$, which can be significantly controlled through nanostructuring. There are several factors that can be monitored, either individually or jointly, to control $\kappa_{\rm ph}$ for nanocomposite materials. These include: inter-atomic bond length relaxation; new features in phonon dispersion curves, changes in phonon velocity; phonon-interface scattering effects, alteration in anharmonic phonon interactions due to the presence of more than one constituent material and larger unit cell size than in bulk materials, and the anisotropic nature of thermal conductivity. Two of these, *viz* phonon-interface scattering and anharmonic phonon interactions, are the most critical factors, but have been studied the least.

In this work we discuss a theory of phonon scattering rates resulting from mass smudging across interfaces and from anharmonicity in nanocomposite materials. We apply our formulations to calculate $\kappa_{\rm ph}$ of ultra-short period Si/Ge[001] superlattices within the single-mode relaxation time approximation for the solution of the phonon Boltzmann transport equation.

2. Theoretical Details

We consider the superlattice structure $A_{n_1}B_{n_2}[001]$, grown along [001], with each unit cell containing n_1 bilayers of A and n_2 bilayers of B. Within the single-mode relaxation time scheme,

[1] Present address: School of Physics, University of Durham, Durham DH1 3LE, UK

the phonon (lattice) thermal conductivity can be expressed as

$$\kappa_{\mu\nu} = \frac{\hbar^2}{N_0 \Omega k_{\rm B} T^2} \sum_{\boldsymbol{q}s} \omega^2(\boldsymbol{q}s) c_{s,\mu}(\boldsymbol{q}) c_{s,\nu}(\boldsymbol{q}) \tau(\boldsymbol{q}s) \bar{n}_{\boldsymbol{q}s}(\bar{n}_{\boldsymbol{q}s}+1), \tag{1}$$

where N_0 is the number of unit cells, Ω the volume of each cell, $\mathbf{q}$ and s label the wave-vector and polarisation of a phonon mode, and $\omega(\boldsymbol{q}s)$, $c_{s,\mu}(\boldsymbol{q})$ and $\bar{n}_{\boldsymbol{q}s}$ are the frequency of a given mode, its $\mu^{\rm th}$ velocity component and the associated value of the Bose-Einstein distribution, respectively. The single mode relaxtion time $\tau(\boldsymbol{q}s)$ is determined by $\tau(\boldsymbol{q}s)^{-1} = \tau_{\rm bs}^{-1} + \tau_{\rm md}^{-1} + \tau_{\rm ep}^{-1} + \tau_{\rm ims}^{-1} + \tau_{\rm anh}^{-1}$. The first four terms are the boundary scattering contrubution, the mass-defect scattering contribution, the electron-phonon contribution and the interface mixing contribution that arises due to imperfections in superlattice interfaces. These are routinely discussed in detail when dealing with $\kappa_{\rm ph}$ for bulk materials. In a recent publication [1] we have provided a formalism for phonon scattering due to anharmonicity and from mass smudging at superlattice interfaces. Here we shall discuss the anharmonic interaction, detailing the derivation of the so-called 'dual mass term' (also used in earlier formulations of the anharmonic scattering [2])

$$\begin{aligned}
\tau_{\rm anh}^{-1}(\boldsymbol{q}s) &= \frac{\pi\hbar\bar{\gamma}^2}{\rho_0 N_0 \Omega \bar{c}^2} \sum_{\boldsymbol{q}'s',\boldsymbol{q}''s'',\boldsymbol{G}} \frac{\left(\mathcal{R}_{\boldsymbol{q}s,\boldsymbol{q}'s',\boldsymbol{q}''s''}\right)^2}{\omega(\boldsymbol{q}s)\omega(\boldsymbol{q}'s')\omega(\boldsymbol{q}''s'')} \mathcal{D}(\boldsymbol{q},\boldsymbol{q}',\boldsymbol{q}'') \\
&\times \left[\frac{\bar{n}_{\boldsymbol{q}'s'}(\bar{n}_{\boldsymbol{q}''s''}+1)}{(\bar{n}_{\boldsymbol{q}s}+1)} \delta(\omega(\boldsymbol{q}s)+\omega(\boldsymbol{q}'s')-\omega(\boldsymbol{q}''s''))\delta_{\boldsymbol{q}+\boldsymbol{q}',\boldsymbol{q}''+\boldsymbol{G}} \right. \\
&+ \left. \frac{1}{2}\frac{\bar{n}_{\boldsymbol{q}'s'}\bar{n}_{\boldsymbol{q}''s''}}{\bar{n}_{\boldsymbol{q}s}} \delta(\omega(\boldsymbol{q}s)-\omega(\boldsymbol{q}'s')-\omega(\boldsymbol{q}''s''))\delta_{\boldsymbol{q}+\boldsymbol{G},\boldsymbol{q}'+\boldsymbol{q}''} \right],
\end{aligned} \tag{2}$$

where $\bar{\gamma}$ is a mode-averaged Grüneissen constant, $\boldsymbol{G}$ is a reciprocal lattice vector, ρ_0 is the average density, $\bar{c}$ is the average acoustic velocity and

$$\mathcal{R}_{i,j,k} = \left[\sqrt{\omega(i)\omega(j)}(\omega(i)+\omega(j))|\omega_\Gamma(k)-\omega(k)| + \text{similar terms with } i,\ j \text{ and } k \text{ interchanged}\right]/3!, \tag{3}$$

where i, j and k label modes and $\omega_\Gamma(i)$ is the frequency of mode i at the Brillouin zone centre. The 'dual mass term' $\mathcal{D}(\boldsymbol{q},\boldsymbol{q}',\boldsymbol{q}'')$ accounts for the phase correlation of the anharmonic force constants within a unit cell of the superlattice structure.

In order to derive a simple and usable form for $\mathcal{D}$, we need to examine the structure of the scattering matrix element for bulk materials, as given in [3]. Typically, for an alloy we use the ansatz [4]

$$\frac{\overline{|A^{ss's''}_{\mathbf{q}\mathbf{q}'\mathbf{q}''}|^2}}{\rho_0} = \frac{4\rho_0^2\gamma^2 c_s^2(\mathbf{q})c_{s'}^2(\mathbf{q}')c_{s''}^2(\mathbf{q}'')}{\bar{c}^2\rho_0^3}, \tag{4}$$

where $A^{ss's''}_{\mathbf{q}\mathbf{q}'\mathbf{q}''}$ is the matrix element for the appropriate polarisations and momenta; ρ_0 is the *average* density. Essentially, we are making the assumption that:

$$\frac{\overline{|A^{ss's''}_{\mathbf{q}\mathbf{q}'\mathbf{q}''}|}}{(\rho\rho'\rho'')} \propto (ee'e'')_{\rm CNT}\rho_0, \tag{5}$$

where $(ee'e'')_{\rm CNT}$ is some number which parameterises the overall behaviour of the eigenvector of each mode in the continuum limit. This simple assumption is justified in alloy systems, where effects arising from the random locations of different species of atoms may be assumed to average out, but it is not so clear that it is justified in ordered systems such as superlattices.

For a superlattice structure $A_{n_1}B_{n_2}$, the atomic layers in sections A and B are treated as patches of densities ρ_1 and ρ_2, respectively. Using Eqs. (4.83) to (4.85) in Ref. [3] as the basis of what follows, we define for each patch a within a period tensor $(A_{ijk}^{lmn})_a$ related to the third order elastic constants; the average of this over a superlattice period is $\overline{A_{ijk}^{lmn}}$. The atomic eigenvectors are approximated as their amplitudes and summation over different directions is neglected (since for superlattices the relevent physics occurs only in one direction). We define $\mathcal{E}(\mathbf{q},q',q'')$ to be a quantity with dimensions of $\rho^{3/2}$ and take $A_{\mathbf{qq'q''}}^{ss's''}$ to be the following product of averages:

$$\begin{aligned} A_{\mathbf{qq'q''}}^{ss's''}\mathcal{E}(\mathbf{q},q',q'') &= \frac{\overline{A_{ijk}^{lmn}}}{(n_1+n_2)^3}\sum_{a,\,bc=1}^{n_1+n_2}\frac{e_a e'_b e''_c}{\sqrt{\rho_a\rho_b\rho_c}} \\ &= \frac{\overline{A_{ijk}^{lmn}}}{(n_1+n_2)^3}\left(\frac{n_1e_1}{\rho_1^{1/2}}+\frac{n_2e_2}{\rho_2^{1/2}}\right)\left(\frac{n_1e'_1}{\rho_1^{1/2}}+\frac{n_2e'_2}{\rho_2^{1/2}}\right)\left(\frac{n_1e''_1}{\rho_1^{1/2}}+\frac{n_2e''_2}{\rho_2^{1/2}}\right), \end{aligned} \tag{6}$$

where e_a is the amplitude of the eigenvector associted with patch a for momentum $\boldsymbol{q}$ and band s, e'_b is the amplitude of the eigenvector associted with patch b for momentum q' and band s' and so on. The above consideration has similarities to the work in [5].

We want to express $A_{\mathbf{qq'q''}}^{ss's''}\mathcal{E}(\mathbf{q},q',q'')$ in terms of ratios of e_1 and e_2 and some average of $e_1e'_1e''_1$ and $e_2e'_2e''_2$. Defining

$$\mathcal{A}_{ij} = \frac{\Gamma_i\rho_i^{3/2}}{e_ie'_ie''_i}\left(\frac{n_1e_1}{\rho_1^{1/2}}+\frac{n_2e_2}{\rho_2^{1/2}}\right)\left(\frac{n_1e'_1}{\rho_1^{1/2}}+\frac{n_2e'_2}{\rho_2^{1/2}}\right)\left(\frac{n_1e''_1}{\rho_1^{1/2}}+\frac{n_2e''_2}{\rho_2^{1/2}}\right), \tag{7}$$

where Γ_1 is the relative proportion of species 1 and $\Gamma_2 = 1-\Gamma_1$ is the relative proportion of species 2, and replacing $e_1e'_1e''_1$ and $e_2e'_2e''_2$ in the denominator with a single mode-averaged parameter e_{over} (we assume that the deviations from this average arising from the effects of superlattice structuring are contained within the parameters $\mathcal{A}_{ij}/\rho_i$) we have:

$$A_{\mathbf{qq'q''}}^{ss's''}\mathcal{E}(\mathbf{q},q',q'') = \frac{\overline{A_{ijk}^{lmn}}e_{\text{over}}}{(n_1+n_2)^3}\left(\frac{\mathcal{A}_{12}}{\rho_1^{3/2}}+\frac{\mathcal{A}_{21}}{\rho_2^{3/2}}\right). \tag{8}$$

Comparing this with the standard expression for anharmonic continuum [3], we can write:

$$|\overline{A_{\mathbf{qq'q''}}^{ss's''}}|^2\mathcal{D}(\mathbf{q},q',q'') = \frac{4\rho_0\gamma^2c_s^2c_{s'}^2c_{s''}^2}{\overline{c}^2}\mathcal{D}(\mathbf{q},q',q''), \tag{9}$$

where the prefactor is assumed to approximate the behaviour of the square of $\overline{A_{ijk}^{lmn}}e_{\text{over}}$ and $\mathcal{D}(\mathbf{q},q',q'') = (\mathcal{E}(\mathbf{q},q',q''))^2$ is the dual mass term, given by

$$\mathcal{D}(\mathbf{q},q',q'') = \frac{1}{(n_1+n_2)^6}\left(\frac{\mathcal{A}_{12}}{\rho_1^{3/2}}+\frac{\mathcal{A}_{21}}{\rho_2^{3/2}}\right)^2. \tag{10}$$

Following previous work [5, 2, 1], we calculate $\mathcal{A}_{ij}$ using an expression that describes the ratio of amplitudes in the diatomic linear chain model.

3. Review of Some Recent Results

Calculations of κ_{ph} of ultra-thin $\text{Si}_n\text{Ge}_n[001]$ superlattices, with each unit cell containing n atomic bi-layers, were carried out in [1]. We consider $\mathcal{P}$ as the probability of the interchange of

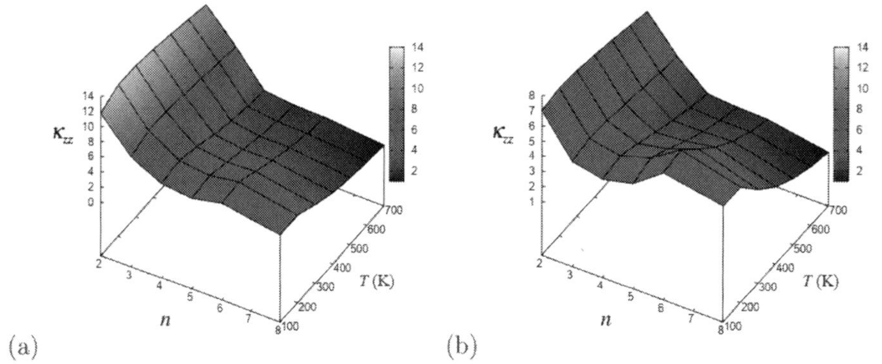

Figure 1. Variation of κ_{ph} (in units of W m^{-1} K^{-1}) along the growth direction in the $Si_nGe_n[001]$ superlattices for: (a) with a fixed interface mass mixing ($\mathcal{P} = 0.05$), and (b) with $\mathcal{P}$ varying with n. The figures are taken from [1].

a number of pairs of bilayers, one from one half of Si and one from the neighboring half of Ge. Figure 1(a) displays the behaviour of κ_{ph} along the growth direction for a number of superlattice periods and temperatures with a fixed amount of interface mass-smudging, which is expressed in terms of the constant probability factor $\mathcal{P} = 0.05$. Figure 1(b) displays κ_{ph} when the interface mass-smudging parameter $\mathcal{P}$ is allowed to vary according to a simple model $\mathcal{P} = 0.24/(n-1)$ of how the degree of mixing should vary with the number of bilayers n in the superlattice. One notable feature of these results is that for fixed $\mathcal{P}$, there is a decrease in κ_{ph} as n is increased. For varying $\mathcal{P}$, we see that κ_{ph} has a minimum at $n = 4$ for temperatures in the vicinity of 100 K, increasing rapidly as n is reduced and more slowly as n is increased, possibly with some degree of oscillation.

4. Summary

We have summarised the derivation of the dual mass term that appears in the anharmonic phonon interaction for superlattice systems. Using this, we have discussed the phonon conductivity results for ultra-thin $Si_nGe_n[001]$ superlattices with two different scenarios for mass-smudging across interfaces. For decreasing probability of mass-smudging away from interfaces, we find that κ_{ph} attains a minimum value for a number of bilayers of around $n = 4$.

Acknowledgements

I. O. T. was supported by EPSRC (UK) Grant Award No. EP/H046690/1 while carrying out this work. The calculations in this work were carried out using the Intel Nehelam (i7) cluster (CERES) at the University of Exeter.

References

[1] Thomas I O and Srivastava G P 2013 *Phys. Rev. B* **88** 115207
[2] Hepplestone S P and Srivastave G P 2011 *Phys. Rev. B* **84** 115326
[3] Srivastava G P *The Physics of Phonons* (Adam Hilger, IOP Publishing Ltd, Bristol, 1990)
[4] Thomas I O and Srivastava G P 2012 *Phys. Rev. B* **86** 045205
[5] Ren S Y and Dow J D *Phys. Rev. B* **25** 3750

Evidence of Longitudinal Acoustic Phonon Generation in Si Doping Superlattices by Ge Prism-Coupled THz Laser Radiation

T Wilson[1*], E Kasper[2], M Oehme[2], J Schulze[2], and K Korolev[1]
[1] Department of Physics, Marshall University, One John Marshall Drive, Huntington, WV 25755 USA
[2] Institut für Halbleitertechnik (IHT), Universität Stuttgart, Pfaffenwaldring 47 70569 Stuttgart, Germany

E-mail*: wilsont@marshall.edu

Abstract. We report on the direct excitation of 246 GHz longitudinal acoustic phonons in silicon doping superlattices by the resonant absorption of nanosecond-pulsed far-infrared laser radiation of the same frequency. A longitudinally polarized evanescent laser light field is coupled to the superlattice through a germanium prism providing total internal reflection at the superlattice interface. The ballistic phonon signal is detected by a superconducting aluminum bolometer. The sample is immersed in low-temperature liquid helium.

1. Introduction

The optical generation of coherent, monochromatic acoustic phonons has been a long-standing goal in solid-state physics. It is has been demonstrated that it is possible to generate coherent pulses of longitudinal acoustic (LA) phonons in a gallium arsenide/aluminium arsenide compositional superlattice by femtosecond laser excitation when the excitation energy matches the E1-HH1 transition in the superlattice [1]. This excitation mechanism used to generate the coherent longitudinal acoustic pulse undoubtedly also generates a burst of incoherent phonons by carrier relaxation in the gallium arsenide. Alternatively, the possibility of direct action of the electric field of a FIR light wave on the layers of bound space charge to resonantly generate single polarization sound, was discussed in the context of doped superlattices, as early as 1983 [2]. In this vein, Ruden *et al.* [3] have studied the effects on the lattice dynamical properties of an otherwise homogeneous semiconductor, stemming from impurity charges associated with a microstructured one-dimensional periodic *n* and *p*-doping sequence, a so-called *nipi* doping superlattice (DSL). The phonon dispersion relation is nearly unchanged, as one might expect from a concentration of 10^{-3} impurity atoms or even less. The periodic space charge resulting from the doping superstructure, however, is the origin of a unique feature: LA or transverse acoustic (TA) phonons, with wavevectors corresponding to odd multiples of the smallest reciprocal superlattice vector, may be directly excited by electromagnetic fields of the corresponding frequency with reasonable conversion efficiency. The fundamental resonant frequency f for the generated phonons is given by $f=s/d$, where s and d are the phase velocity for LA phonons and the superlattice period, respectively.

[1*] To whom any correspondence should be addressed.

2. Theory

LA acoustic phonons couple to electric fields normal to the DSL layers. The *conversion efficiency* ξ is defined as the ratio of the generated acoustic energy to the energy of the incident electromagnetic wave and is given by equation (1) (cgs units) below, where a is the host lattice constant (assumed simple cubic lattice), $n_D^{(2)}$ is the two-dimensional doping concentration, κ_0 is the dielectric constant, M is the atomic mass, d is the superlattice period, c is the speed of light, s is the phonon phase velocity, N is the number of superlattice periods, and $r \equiv (d/a)$, is the enlarged unit cell:

$$\xi = \left[\frac{4\pi d(ea^2 n_D^{(2)} N)^2}{\kappa_0^{3/2} Mrcs} \right] \quad (1)$$

The upper limit to the two-dimensional (donor or acceptor) doping concentration is determined by the condition that there be an effective bandgap. The limiting doping concentration follows from equations (1) and (8) in Ruden [3], and is given in equation (2) below:

$$n_D^{(2)} = \frac{2E_g \kappa_0}{\pi d e^2} \quad (2)$$

where E_g is the host bandgap. At this doping level, the thin sheets of acceptors and donors are degenerate and there will be no carrier freeze out at liquid helium temperatures. In its ground state, the sample thus consists of alternately positively and negatively charged sheets of ions. Care must be taken to insure that the sample has a closely balanced dopant concentration.

3. Superlattice and THz laser source

We use a delta-doped DSL consisting of 30 periods, grown on a [100] float zone high-resistivity 0.61 mm thick silicon substrate. The superlattice layers have a period d = 34.5 nm and the individual layers have a two-dimensional doping concentration $n_D^{(2)} = 8.5 \times 10^{12}$ cm^{-2}. From equation (1), the conversion efficiency of this DSL is $\xi = 9.6 \times 10^{-9}$. The period was chosen to match the LA phonon wavelength in silicon at a frequency (246 GHz) corresponding to a particularly strong FIR laser line (1.22 mm) produced with a $^{13}CH_3F$ gain medium. An experimental requirement was the development of a nanosecond pulsed kilowatt FIR laser source - kilowatt power is required for the relatively small conversion efficiency of electromagnetic power into acoustic power, and nanosecond pulses are required to resolve the LA and TA phonon time-of-flights across the thin substrates typically used for MBE-based DLS growth. The author has previously described his development of a cavity-dumped multi-kilowatt source of nanosecond-pulsed FIR laser radiation by optically-switching a Brewster-angle mounted silicon wafer placed within a conventional pulsed optically-pumped molecular gas laser [4]. In this experiment, radiation from the 9P32 TEA CO_2 laser line, containing 300 mJ in a 300 ns pulse width at 10 prf, is hole-coupled into the FIR laser resonator. The FIR resonator is a near-concentric design with stability parameters g_1 and g_2 equal to -0.12 and -0.52, respectively. The smallest beam waist w_0 is *12.6 mm* at the silicon wafer. The cavity-dumped FIR laser radiation output is then mode-matched, using TPX lenses, into a 246 GHz overmoded H11 segmented corrugated waveguide of inner diameter 22 mm and length 3 m. A Teflon lens at the end of the waveguide then focuses the radiation to a beam diameter of 4 mm upon at the Ge prism-coupled DSL. The typical FIR laser pulse width is 6-7 ns and corresponds to the cavity-dumping time for the round-trip distance (2 m) between the silicon wafer and one end resonator mirror. The average pulse energy is measured with a *Scientech* calorimeter and corresponds to a peak power of ~10 kW. The Q-switched YAG laser pulse may be delayed relative to the TEA CO_2 pump pulse by a delay generator such that a longer delay results in decreased FIR pulsed power.

4. Experiment

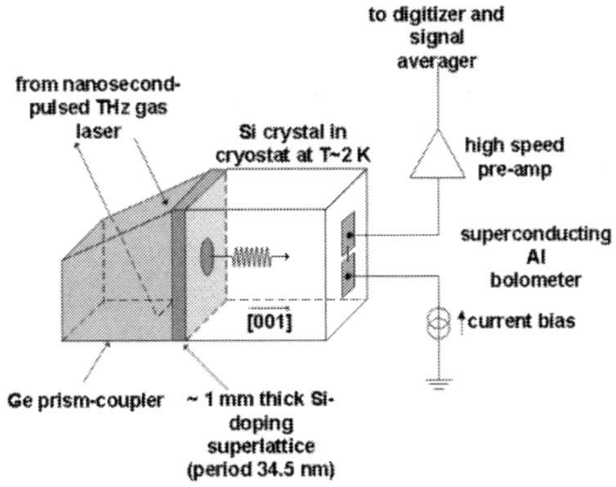

Figure 1. Schematic of experimental arrangement.

The experimental arrangement is shown above in figure 1. 246 GHz p-polarized laser radiation is transported via the corrugated waveguide and focused through a Teflon lens, enters through a crystal quartz window of a *Janis Research* 8-CNDT immersion cryostat with optical access and illuminates a Ge prism in contact with the DSL. The DSL/Si substrate has a frontal area of 5 mm x 7 mm and is 0.61 mm thick. The THz laser radiation is incident upon a truncated 30-60-90 degree high-purity Ge prism. The prism base is of dimension 9.2 mm x 6.1 mm. The incident angle of 60 degrees at the Ge/Si DSL interface exceeds the critical angle (58.4 degrees) and the FIR light is totally internally reflected. The resulting evanescent field within the superlattice is longitudinally-polarized normal to the superlattice layers with a decay length at this frequency of 0.35 mm. Phonon detection is made with a superconducting bilayer granular aluminum/palladium bolometer operating in a constant-current mode. We fabricate the bolometers using a bilayer lift-off recipe with sputtering, as described previously [5]. The bolometer is of dimension 10 x 20 μm^2 and 100 nm thick. The bolometer responsivity is calculated from characterization to be 1.7 kV/W. Typical bias temperatures and currents are 1.8 K and 20 μA, respectively. Temperature control is maintained by pumping on the vapor above the liquid helium bath through an *MKS 653B* throttle valve with an *Oerlikon Sogevac SV-100B* vacuum pump. The throttle valve aperture is adjusted under *LabView* control with an MKS 600 pressure controller with feedback from an *MKS 626* pressure sensor. Temperature stability is better than 0.2 mK near 1.8 K.

5. Results and Discussion

Figure 2 shows first results of the presence of the expected coherent acoustic phonon signal. The amplified bolometer response (in mV) is plotted versus time (in nsec). Three traces are shown, corresponding to different levels of bolometer bias current. The bolometer signal also contains a lower-frequency noise component due to the pickup of RFI from the TEA CO_2 laser that is manifested here by a sloped baseline that is also a function of bias current. The cavity-dumping time delay also has been increased from the optimal delay in order to minimize the RFI noise signal. The first pulse observed, occurring near 280 ns in figure 1, is assumed to be the decaying evanescent THz light field that leaks into the substrate and illuminates the bolometer. It is coincident with the YAG Q-switching 532 nm light pulse (not shown). The second pulse, arriving ~50 ns later, is assumed to be the coherent longitudinal acoustic phonon pulse traveling ballistically across the 0.61 mm thick substrate. If one uses the LA sound speed in [100] silicon of 8450 m/s obtained from low-frequency acoustic velocity measurements [6], the expected arrival time would be considerably longer however, namely, 72 ns. The shorter time-of-flight of the measured phonon pulse could result from the elastic constants at 246 GHz being substantially different, or from a stiffening of the silicon due to stress caused by the phonons. The latter would then probably be related to solitons [7]. We also note the absence of any observed delayed and larger TA phonon pulse as occurs with heat pulse production along [100] Si.

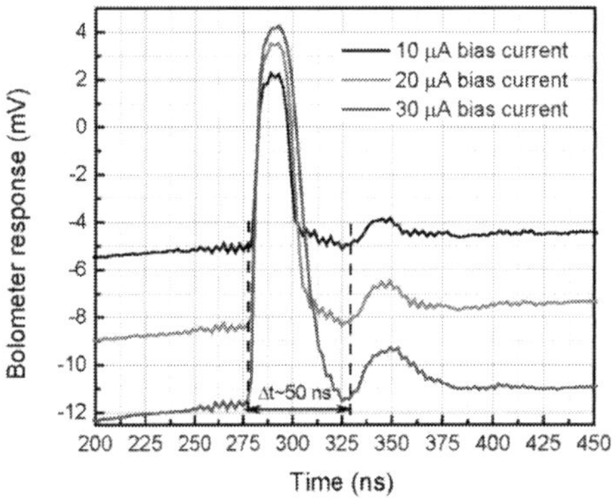

Figure 2. Bolometer signals that show the presence of a secondary pulse (near 325 ns) presumed to be due to coherent phonons arriving at a delay of 50 ns from the earlier incident laser light pulse (near 275 ns).

6. Conclusions and Future Work

We have presented evidence for coherent acoustic phonon generation by the resonant absorption of Ge prism-coupled nanosecond-pulsed 246 GHz FIR laser radiation in silicon delta-doped doping superlattices. The technique produces *single polarization* acoustic phonons with negligible associated broadband heat pulse production. The time-of-flight for the ballistic phonon pulse observed in [100] Si is significantly smaller than that expected for lower-frequency sound speeds. We will soon use two metal-mesh polarizers to rotate the polarization of the incident FIR laser radiation for an in-plane electric field polarization to see if the phonon signal is reduced or extinguished as expected. Future plans are to: (1) include the use of a faster (30-nm thick) bolometer to investigate the presence of acoustic solitons on the front edge of the phonon signal, (2) conduct similar investigations at 1.04 THz, (3) use Si:B piezophonon spectroscopy [8] to verify the monochromaticity of the phonon signal and (4) use a stack of DLS's each with separate resonant frequencies to realize a discretely tunable coherent THz acoustic phonon source.

Acknowledgements

We acknowledge funding from NASA EPSCoR Grant No. NNX11AM04A, AFOSR Grant No. FA9550-12-1-0100, West Virginia Higher Education Policy Commission Grant No. HECP.DSR.12.129, and the use and staff support of both the West Virginia University Shared Resource Facility and the Marshall University Molecular and Biological Imaging Center.

References

[1] Kent A J, Stanton N M, Challis L J, and Henini M 2002 *Appl. Phys. Lett.* **81** (18), 3497
[2] Quinn J J, Strom U, and Chang L L 1983 *Solid State Communications* **45** (1) 111
[3] Ruden P, and Dohler G H 1983 *Solid State Comm.* 45 (1) 23
[4] Wilson T E 1999 *Phys. Rev. B* **59** (20), 12996
[5] Wilson T E 2011 *Meas. Sci. Technol.* **18**, N53
[6] Goto T, Yamada-Kaneta H, Saito Y, Nemoto Y, Sato K, Kakimoto K and Nakamura S 2006 J. Phys. Soc. Jpn. **75** 044602
[7] Kent A J, and Stanton N M 2007 *J. Phys.: Conf. Series* **92**, 012004
[8] Schwarte M, and Berberich P 1985 *J. Phys. C: Solid State Physics* **18,** 3225

IOP Conf. Series: Materials Science and Engineering **68** (2014) 012009 doi:10.1088/1757-899X/68/1/012009

Numerical calculation of optical phonon decay rate in InN/GaN MQW

H Xia*, R Patterson, Y Feng, T Smyth, Y Liao, P Zhang, X Dai, N Gupta, X Wen, S Chung, X Jia, L Wu, Z Lin, B Puthen-Veettil, S Huang, S Shrestha, G Conibeer

Australian Centre for Advanced Photovoltaics, University of New South Wales, Sydney, Australia

E-mail: *h.xia@unsw.edu.au

Abstract. Anharmonic decay of high frequency phonons into low frequency vibrations is a significant energy loss mechanism in semiconductors. In the field of Hot Carrier Solar Cells (HCSC), preventing this decay is of great importance as it helps increase hot carriers lifetime. Phonon decay in nitride compounds as well as their consisting nano-crystals like quantum dots and multiple quantum wells (MQWs) has not been extensively studied in the literature. In this work, the decay channels of the A1 and the high-lying E2 optical phonons in an InN/GaN MQW are analysed. We find that the no Klemens decay is present in A1 mode whereas on the contrary E2 is dominated by this process. We also observe that the linewidth for A1 is enlarged a few times than the bulk counterpart while that for E2 remains similar, which is attributed to different vibration nature.

1. Introduction

Nano-crystals or nano-materials have been a hot topic since the 1990s when people started to investigate physical phenomenon in such a small scale. Since the debut of nano-crystal prototypes, interpreting the physics behind the findings became inevitable. Due to the confinement effects, transit from bulk to nano-scale properties has attracted researchers' interests, not only because the physics is not clear but also because the differences nano and bulk materials post a wide range of potential applications. A great number of models have been applied to the field of nano-crystals, such as nano-tubes, nano-rods, quantum dots and quantum wells, so that novel properties can be determined even before materials and devices are fabricated. In this paper, a numerical model is built to investigate the optical phonon decay rate in both bulk nitride compounds and their counterpart MQW structure. This study aims to unravel whether MQWs are suitable for HCSCs in terms of phonon lifetime.

Long hot carrier lifetime is the most essential property of HCSC, under which HCSC can reach a very high energy conversion efficiency [1] far beyond the Shockley-Queisser limit [2]. The requirement under which a HCSC can differ from a conventional solar cell is actually quite strict. Optical phonon lifetime of around a few hundred pico-seconds is required but normal bulk materials only have lifetime of a few to some tens pico-seconds or less. One potential mechanism of blocking phonon decay was proposed in the prototype HCSC model. It states that if the phononic band-gap between the optical and the acoustic phonon branch is larger than the energy of the highest acoustic phonon, the optical phonon lifetime could be increased [3]. This is based on an assumption that the Klemens decay [4] is the dominant channel in materials. In this research, this assumption is examined in an InN/GaN MQW consisting of two atomic layers of both InN and GaN, which is also hot carrier absorber candidate, owing to its multi-layer structure and large mass atomic ratio [5].

2. Anharmonic theory

$$\Gamma = \frac{2\pi}{\hbar} \left|\langle\psi_i|H|\psi_f\rangle\right|^2 \tag{1}$$

In an anharmonic crystal, phonon decay is inevitable and in most circumstances three-phonon process is the dominant mechanism because the probability for four-phonon or even higher processes drops extremely fast. The Klemens mechanism states that a phonon can decay into two acoustic phonons of equal energy and opposite momentum. In general, as long as energy and momentum are conserved the decay process can occur. Fortunately, the Fermi's golden rule allows one to evaluate the transition rate using a simple and elegant formula [EQ 1]. Different transitions happen at different rates depending on the coupling between the initial state and the final state characterised by the Hamiltonian. In the case of phonon decay, the Hamiltonian contains phonon creation and annihilation operators as well as the perturbed crystal energy. The initial state is usually the state of interest, i.e. optical phonons, while the final states in this situation are the lower-branch phonons. In this study, we only consider zone-centre optical phonon decay processes including both Klemens' and non-Klemens Channel as this kind of phonon is most likely to interact with photons.

$$\begin{aligned} E_{tot}(\mathbf{u}) = E_{tot}^{(0)} &+ \frac{1}{2} \sum_{ll',ss',\alpha\beta} \frac{\partial^2 E_{tot}}{\partial u^l_{s,\alpha} \partial u^{l'}_{s',\beta}} u^l_{s,\alpha} u^{l'}_{s',\beta} \\ &+ \frac{1}{6} \sum_{ll'l'',ss's'',\alpha\beta\gamma} \frac{\partial^3 E_{tot}}{\partial u^l_{s,\alpha} \partial u^{l'}_{s',\beta} \partial u^{l''}_{s'',\gamma}} u^l_{s,\alpha} u^{l'}_{s',\beta} u^{l''}_{s'',\gamma} + \mathcal{O}(\mathbf{u}^3) \end{aligned} \tag{2}$$

The total energy of the crystal can be expanded into Fourier series in terms of atomic displacements or phonons. In EQ 2, $\mathbf{u}$ represents the atomic displacement while the indices s, l and (α β γ) are the atomic species, number of unit cell and the three Cartesian coordinates respectively. The third term in this equation gives the perturbed crystal energy due to the anharmonic three phonon process. In the second quantisation theory, atomic displacement can be translated into reciprocal space and same for the third order crystal energy. After all the algebra, the Hamiltonian operator for the three phonons decay process reads:

$$\begin{aligned} H^{anh} = \frac{1}{6} \sum_{\mathbf{q},\mathbf{q}',\mathbf{q}''} \sum_{j_0,j_2,j_3} U(\mathbf{q},\mathbf{q}',\mathbf{q}''; j_0, j_1, j_2) &\left(a^\dagger_{j_0}(-\mathbf{q}) + a_{j_0}(\mathbf{q})\right) \\ &\left(a^\dagger_{j_1}(-\mathbf{q}) + a_{j_1}(\mathbf{q})\right) \left(a^\dagger_{j_2}(-\mathbf{q}) + a_{j_2}(\mathbf{q})\right) \end{aligned} \tag{3}$$

$a^\dagger$ and a are simply the phonon creation and annihilation operators while $U(\mathbf{q},\mathbf{q}',\mathbf{q}''; j_0, j_1, j_2)$ in EQ 3 denotes the Fourier-transformed anharmonic crystal energy which is evaluated within the Density Functional Theorem (DFT) framework in this study. U is the most important factor in the calculation. Without a well established method, any result from the Fermi's golden rule is nothing more than a rough estimation. In the past, a single macro coupling constant was usually used for all possible transitions, which we are still able to meet in many recent studies. Evaluating the constants by DFT was first attempted by Debernardi [6] in the 1990s. This has proven to be a powerful and accessible method as modern computing capacity continues its rapid growth. Computing the coupling constants is made possible by the so-called $2n+1$ theorem [7] which allows one to go to the nth order crystal energy at minimum cost. In this study, all physical quantities including lattice parameters, phonon dispersions and the anharmonic terms were extracted from the results produced by the package "Quantum Espresso" [8].

We then apply the Fermi's golden rule, simplifying the expression by using the properties from phonon creation and annihilation operators: $a^\dagger|n\rangle = \sqrt{n+1}|n+1\rangle$ and $a|n\rangle = \sqrt{n}|n-1\rangle$, where n is the Bose-Einstein occupation. Eventually one reaches the final formula of zone centre optical phonon decay rate via all possible channels:

$$\Gamma_{LTO} = \frac{\pi}{2\hbar}\sum_{\mathbf{q},j_1,j_2} \left|U(\mathbf{0},\mathbf{q}',\mathbf{q}'';LTO,j_1,j_2)\right|^2 \{[n_{j_1}(\mathbf{q}) + n_{j_2}(\mathbf{q}) + 1]\delta[\omega_{LTO}(\mathbf{0}) - \omega_{j_1}(\mathbf{q}) - \omega_{j_2}(-\mathbf{q})] + |n_{j_1}(\mathbf{q}) - n_{j_2}(\mathbf{q})|\delta[\omega_{LTO}(\mathbf{0}) \pm \omega_{j_1}(\mathbf{q}) \mp \omega_{j_2}(-\mathbf{q})]\} \quad (4)$$

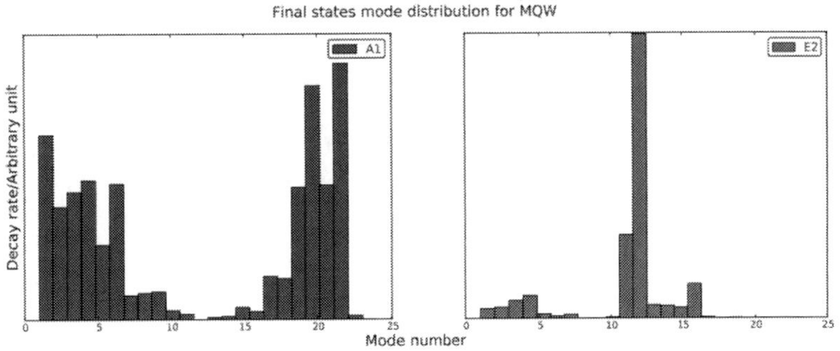

Figure 1. Mode-resolved final state distribution for the MQW. Mode number is simply the number of the eigenstate. A1 is at mode number 24 while the E2 is at 20. In this particular structure, we have 24 modes in total for one reciprocal point.

Table 1. Decay channel comparison and predicted linewidths for A1 and E2 modes. The two numbers under the channel columns are the channel occupation and the actual contribution to the decay rate respectively. The linewidth and the lifetime are predicted at $T = 10$ K.

	Klemens[%]	Ridley[%]	Others[%]	linewidth [cm^{-1}]	lifetime [ps]
A1	0.00, 0.00	29.5, 48.9	70.5, 51.1	55.30	0.10
E2	10.4, 55.2	55.9, 15.1	33.8, 29.8	1.59	3.34

3. Decay channel analysis and conclusion

We observe that the decay channels in the MQW studied in this work are rather complicated consisting of not only Klemens and Ridley [9] channels, but also other pathways (Figure 1 and Table 1). The Ridley channel represents decay into one lower energy optical phonon and one acoustic phonon, while the other channels contain all non-Klemens and non-Ridley pathways. The contributions to the final decay rates for different decay channels are different, of which the assumed dominant Klemens channel appears to be comparable with Ridley channel and other channels. In Table 1, one can see that there is no Klemens channel for A1 mode however the Ridley and other channels make up the whole decay rate. The case in E1 mode is quite different. The Klemens channel only occupies 10.4% of possible paths but is responsible for 55.2% of the total decay rate. In contrast, Ridley channel consists of more than half of the possible paths but only contributes 15.1% to the total rate. Even the other channels contribute more than the Ridley channel does.

Consistency is demonstrated between the computed decay rates and Raman experiments [10, 11, 12]. In the latter, the authors suggested that in general, A1 mode transits to lower energy phonons at a much faster rate than E2 mode. They reported that the linewidths for A1 and E2 mode in GaN are >11.0 and 1.67~3.9 cm^{-1} while the widths in InN are 15.0 and 2.0~5.7 cm^{-1}. This is due to the fact that the phonon energy for A1 mode is some tens to hundred cm^{-1} higher than that for E2 mode, which naturally allows more decaying paths. Compared with the bulk materials, the decay rate for the A1 mode in this MQW is a few times faster than that in the bulk. The E2 mode seems to be comparable. This is reasonable because the A1 mode is the vibration along the confined direction in which the MQW is grown. As the MQW structure reshapes the crystal structure along that direction and more permitted modes exist, A1 can easily decay into modes that conserve both energy and momentum. Contrarily, for E2 mode vibrating in a transverse way, the difference between in bulk and in MQW is limited.

In conclusion, we have examined and analysed possible decay channels in an InN/GaN MQW. Our results show that Ridley and other possible channels can be competitive with the Klemens channel. For phonons vibrating along the growing direction, their lifetime may be suppressed due to more allowed decay paths introduced by the quantum well structure. It seems that our findings do not favour the MQW structure in terms of a longer optical phonon lifetime. However, before drawing this conclusion, it is very important to analyse all other optical modes in the MQW, which could be the subject of future work.

Acknowledgement

This Program has been supported by the Australian Government through the Australian Renewable Energy Agency (ARENA). Responsibility for the views, information or advice expressed herein is not accepted by the Australian Government. This work is also supported by NSW Government Science Leverage Fund.

References

[1] Feng Y, Aliberti P, Veettil B P, Patterson R, Shrestha S, Green M a and Conibeer G 2012 *Appl. Phys. Lett.* **100** 053502 ISSN 00036951
[2] Green M A, Conibeer G, Konig D, Shrestha S, Huang S and et al 2010 *Conf. Rec. {IEEE} Photovolt. Spec. Conf.* (Honolulu, {HI}, United states) pp 57–60
[3] Conibeer G, Green M, Cho E C, König D, Cho Y H, Fangsuwannarak T, Scardera G, Pink E, Huang Y, Puzzer T, Huang S, Song D, Flynn C, Park S, Hao X and Mansfield D 2008 *Thin Solid Films* **516** 6748–6756 ISSN 00406090
[4] Klemens P G 1966 *Phys. Rev.* **148** 845–848
[5] Conibeer G, Patterson R, Huang L, Guillemoles J F, Knig D, Shrestha S and Green M a 2010 *Sol. Energy Mater. Sol. Cells* **94** 1516–1521 ISSN 09270248
[6] Debernardi A, Baroni S and Molinari E 1995 *Phys. Rev. Lett.* **75** 1819–1822 ISSN 0031-9007
[7] Gonze X and Vigneron J P 1989 *Phys. Rev. B* **39** 13120–13128 ISSN 0163-1829
[8] Giannozzi P, Baroni S and et al 2009 *J. Phys. Condens. Matter* **21** 395502 ISSN 0953-8984, 1361-648X
[9] Ridley B 1996 *J. Phys. Condens. Matter* **8** 511–513
[10] Domènech-Amador N, Cuscó R, Artús L, Yamaguchi T and Nanishi Y 2011 *Phys. Rev. B* **83** 245203
[11] Bergman L, Alexson D, Murphy P L, Nemanich R J, Dutta M, Stroscio M A, Balkas C, Shin H and Davis R F 1999 *Phys. Rev. B* **59** 12977–12982
[12] Pu X D, Chen J, Shen W Z, Ogawa H and Guo Q X 2005 *J. Appl. Phys.* **98** –

AUTHOR INDEX

Institute of Physics
Dirac House, Temple Back
Bristol BS1 6BE UK

CURRAN ASSOCIATES INC
proceedings.com